MATERIALS SCIENCE AND TECHNOLOGIES SERIES

SURFACE MODIFIED BIOMEDICAL TITANIUM ALLOYS

MATERIALS SCIENCE AND TECHNOLOGIES SERIES

Magnetic Properties of Solids
Kenneth B. Tamayo (Editor)
2009. ISBN: 978-1-60741-550-3

Mesoporous Materials: Properties, Preparation and Applications
Lynn T. Burness (Editor)
2009. ISBN: 978-1-60741-051-5

Physical Aging of Glasses: The VFT Approach
Jacques Rault
2009. ISBN: 978-1-60741-316-5

Physical Aging of Glasses: The VFT Approach
Jacques Rault
2009. ISBN: 978-1-61668-002-2 (Online Book)

Graphene and Graphite Materials
H. E. Chan (Editor)
2009. ISBN: 978-1-60692-666-6

Dielectric Materials: Introduction, Research and Applications
Ram Naresh Prasad Choudhary and Sunanda Kumari Patri
2009. ISBN: 978-1-60741-039-3

Handbook of Zeolites: Structure, Properties and Applications
T. W. Wong
2009. ISBN: 978-1-60741-046-1

Strength of Materials
Gustavo Mendes and Bruno Lago (Editors)
2009. ISBN: 978-1-60741-500-8

Photoionization of Polyvalent Ions
Doris Möncke and Doris Ehrt
2009. ISBN: 978-1-60741-071-3

Building Materials: Properties, Performance and Applications
Donald N. Cornejo and Jason L. Haro (Editors)
2009. ISBN: 978-1-60741-082-9

Concrete Materials: Properties, Performance and Applications
Jeffrey Thomas Sentowski (Editor)
2009. ISBN: 978-1-60741-250-2

Corrosion Protection: Processes, Management and Technologies
Teodors Kalniņš and Vilhems Gulbis (Editors)
2009. ISBN: 978-1-60741-837-5

Corrosion Protection: Processes, Management and Technologies
Teodors Kalniņš and Vilhems Gulbis (Editors)
2009. ISBN: 978-1-61668-226-2 (Online Book)

Handbook on Borates: Chemistry, Production and Applications
M.P. Chung (Editor)
2010. ISBN: 978-1-60741-822-1

Smart Polymer Materials for Biomedical Applications
Songjun Li , Ashutosh Tiwari, Mani Prabaharan and Santosh Aryal (Editors)
2010. ISBN: 978-1-60876-192-0

Definition of Constants for Piezoceramic Materials
Vladimir A. Akopyan, Arkady Soloviev, Ivan A. Parinov and Sergey N. Shevtsov
2010. ISBN: 978-1-60876-350-4

Handbook of Photocatalysts: Preparation, Structure and Applications
Geri K. Castello (Editor)
2010. ISBN: 978-1-60876-210-1

Organometallic Compounds: Preparation, Structure and Properties

H.F. Chin (Editor)

2010. ISBN: 978-1-60741-917-4

Surface Modified Biomedical Titanium Alloys

Aravind Vadiraj and M. Kamaraj

2010. ISBN: 978-1-60876-581-2

MATERIALS SCIENCE AND TECHNOLOGIES SERIES

SURFACE MODIFIED BIOMEDICAL TITANIUM ALLOYS

ARAVIND VADIRAJ
AND
M. KAMARAJ

Nova Science Publishers, Inc.
New York

For permission to use material from this book please contact us:
Telephone 631-231-7269; Fax 631-231-8175
Web Site: http://www.novapublishers.com

LIBRARY OF CONGRESS CATALOGING-IN-PUBLICATION DATA

Vadiraj, Aravind.
Surface modified biomedical titanium alloys / Aravind Vadiraj, M . Kamaraj.
p. cm.
Includes bibliographical references and index.
ISBN 978-1-60876-581-2 (softcover)
1. Titanium alloys--Biocompatibility. 2. Surface preparation. I. Kamaraj, M . II. Title.
R857.T57V33 2009
610.28'4--dc22 2009045659

Published by Nova Science Publishers, Inc. ✝ New York

Contents

ABSTRACT

Load bearing implants such as hip joints, knee joints and bone plates etc. are prone to failure from the synergistic effect of fretting wear and fatigue during physical bodily movement. Fretting wear or fretting fatigue is a form of adhesive wear phenomenon wherein a small tangential oscillatory motion under high contact pressure gradually erodes the surface and initiates crack within the contact leading to ultimate failure of the material under fatigue loading condition. Modular junctions of hip implants consist of ball on a tapered shaft experiencing fretting wear during body movements.

Titanium alloys, Stainless steels and Co-Cr-Mo alloys are the most commonly used alloys for medical grade devices. Titanium alloys has high strength to weight ratio, superior biocompatibility and corrosion resistance compare to other materials. Surface modified titanium alloys have better tribological properties. In this work, the substrate materials used are Ti-6Al-4V and Ti-6Al-7Nb. PVD TiN coating, plasma nitriding, nitrogen ion implantation, laser nitriding favors formation of TiN and Ti_2N of different thickness according to the process. Thermal oxidation process favors formation of hard and brittle oxide layer.

Fretting wear tests were conducted to study the quality of the coatings and modified layers. Laser nitriding and PVD TiN coating has shown better performance than other coatings due to high hardness of the layers. Friction coefficient for PVD TiN coating is around 0.2 throughout the test. Wear volume for PVD TiN coated and laser nitrided samples were almost 10 and 50 times lesser than other coatings respectively.

Fretting fatigue life of surface modified titanium alloys has considerably improved compared to unmodified materials. Plasma nitrided pairs have shown the best performance over all the coatings. The average fretting fatigue lives of unmodified pairs were 15 to 18% of plasma nitrided pairs, 45 to 50% of the PVD TiN coated pairs and about 60% of ion implanted pairs. Fretting of unmodified alloy pairs have shown high friction and oxidation at the contact due to metallurgical compatibility of the pairs. Fretting of PVD TiN pairs have shown delamination and subsequent oxidation. The damage and friction generated within the contact is a complex interaction between third body particulates, oxide debris and ringer solution. Fretting fatigue life is more for plasma nitride pairs compared to all other modification processes. The damage of ion implanted pairs is similar to unmodified alloys with little improvement in fretting fatigue life. Laser nitrided pairs and thermally oxidation pairs have shown poor fretting fatigue life due to high case thickness and inhomogenities of the layers formed. Friction generated is low compared to al other process, but the specimens experienced premature failure at higher loads.

Chapter 1

INTRODUCTION

1.1. MATERIALS

Materials used for biomedical implants are medical grade stainless steels (SUS 316 LV), Co-Cr-Mo alloys, and titanium alloys (Sumita et. al., 1994). The elements used in these alloys such as Ni, Cr, Mo and Co are not so compatible to the human system (Mark et. al., 1998, Kamachi Mudali et. al. 2003). Titanium alloys are widely used for various applications such as aerospace, chemical industries and biomedical industries. They are selected for biomedical applications due to superior performance in terms of biocompatibility and corrosion resistance within the human body (Mark et. al., 1998). They also have high specific strength, high hardness, and low stiffness compared to other materials which helps in minimizing bone resorption. The primary drawback of titanium based biomedical alloys is its poor resistance to sliding wear or fretting wear which needs immediate attention for increasing the service life as implants. Leyens and peters (2004) attribute the poor wear resistance of pure titanium and titanium alloys to lower c/a ratio of α phase. The wear damage can be minimized by altering the surface composition, which favors the formation of hard ceramic compounds with better tribological properties.

1.2. SURFACE ENGINEERING OF BIOMEDICAL IMPLANTS

Surface engineering is one of the emerging areas in the field of material science or tribology aiming towards modification of the surface for improving its life and efficiency while retaining its original bulk properties. It shows improvement in in-service performance, useful working lifetimes, aesthetic appearance or economics of production. The surface modified components not only improve the tribological performance of the parts but also minimize the investment on repair or replacement. Surface engineering of biomaterials has attracted attention for various important reasons. Wear, fretting and corrosion have become the most important concern in designing the various implants. Harder and biocompatible surfaces of implanted materials prolong the life of the device within the harsh human body environment. Human body fluid contains blood, tissue fluids, amino acids and proteins (Sumita et. al., 1994). Lower partial pressure of oxygen within the body accelerates corrosion of metallic implants due to reduced availability of oxygen for repassivation (Kamachi Mudali et. al., 2003). The various forms of surface damage produce unwanted products which invite hostile tissue reactions and also minimize the life of the implant. In biomedical devices, quality overrules the cost of the product. It is extremely necessary to maintain the superior quality of the device for improved performance in the human body. Therefore surface modification has become an important part of the manufacturing processes for all the bio implants in the recent times.

1.3. Importance of Surface Treatments for Titanium Alloys

Surface modification of titanium alloys are also widely studied to serve the purpose of improved performance within the aggressive body environment. It is observed that surface modified titanium alloys perform much better in terms of both biocompatibility and wear resistance than virgin materials. Elements such a nitrogen, carbon and oxygen are introduced into the surface by various methods thereby altering the chemistry of the surface for superior performance. Nitrogen alloyed titanium is by far the widely studied area for medical applications. It favors formation of hard and chemically resistant titanium nitride/oxides which offers high resistance towards fretting damage. However there is always an optimum level of modified layer thickness beyond which the performance of load bearing components would degrade due to lower ductility and hence strain tolerating capacity of the modified layers.

1.4. FRETTING WEAR AND FRETTING FATIGUE

Fretting phenomenon is observed when contacting bodies are under vibration induced relative sliding (100 - 200μm) (Mutoh, 1995, Waterhouse, 1972). Fretting wear is due to micro motion between contacting bodies induced by vibration and fretting fatigue is due to micro motion induced by cyclic loading of one of the contact member. Fretting wear induces localized damage at the contact, but fretting fatigue causes catastrophic failure of cyclically loaded member. It is common to observe such a thing in flywheel fitted to shaft fabricated as one assembly under cyclic load. Fretting damage is observed along the circumference of such a joint. Bearings of automobiles and gun mountings in military vehicles have also incurred such problems due to vibration during its motion (Waterhouse, 1972). In some cases fretting is considered beneficial to dampen vibration by absorption of the energy. Fretting can also take place in different temperatures and environments depending upon the application of the material. Depending upon the medium, it can take place in sea water, within the human body, nuclear reactors or heat exchangers. Fretting can also produce deleterious effects in electrical contacts, bio implants, steel ropes, cycle pedals and industrial valves. The damages caused by fretting are classified into pits, debris, scratches, metal transfer, surface plasticity, subsurface cracking, and craters. Dobromirski, 1992 indicates that there are more than fifty parameters that govern fretting phenomenon of the interacting surfaces.

The main principle of fretting wear experiments is described in Figure 1.1. It involves two contacting surface under a definite normal pressure. The contact configuration can be of different types. Ball-on-flat arrangement is most commonly used. Fretting wear is induced with definite slip amplitude and normal force (indicated by arrows). Tangential force is monitored with force transducers butted to a stationary member of the contact pair. Data acquisition systems are used to record the various parameters. Fretting wear results are represented as variation of tangential force with time or fretting cycles, changes in force-displacement (F-D) loop with time or number of cycles, scar micrographs, roughness profile along the scar and wear volume generated.

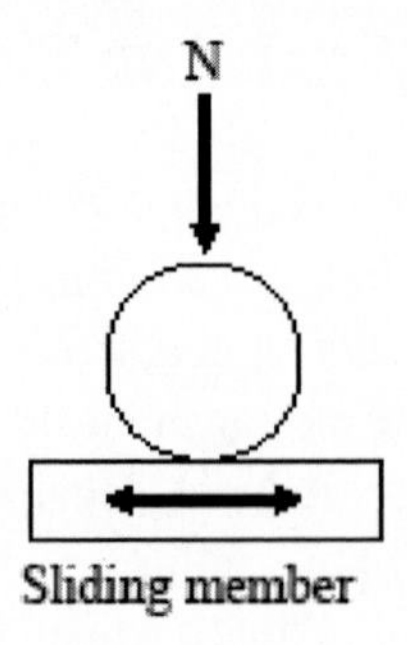

Figure 1.1. Principle of fretting wear.

Fretting fatigue is simulated in the laboratory using tensile specimen and contact pads as shown in Figure 1.2. There are wide varieties of specimens and contact pad configurations that have been used to conduct laboratory scale fretting fatigue experiments. Contact pads are used to apply normal pressure on opposite sides of the tensile specimen subjected to cyclic stress amplitude at certain frequency. The area of contact initially is always less than the real area due to surface irregularity. This is called the apparent area of contact. Crystallographically smooth surface is generally difficult to prepare. Asperities wear down and the area of contact gradually increases with time. When a calibrated normal force (P) is applied on the specimen, tangential force (Q) is generated at the interface during sliding. Extensive localized plastic deformation at the interface area can be observed depending upon the material and magnitude of contact pressure. The relative motion between rubbing interfaces is normally measured as slip amplitude (δ), which is one of the important parameter to predict the life of the components.

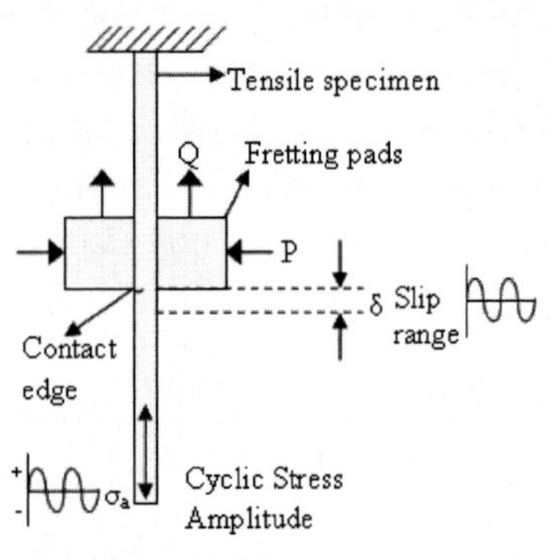

Figure 1.2. Schematic diagram of fretting fatigue phenomena.

Laboratory tests must simulate the original fretting conditions of the contact as closely as possible in order to correlate the results to the original service of the components. Therefore fretting experiments involve large number of independent variables acting synergistically introducing complexity in quantifying and correlating them to the final life of the components. Localized fretting action will also raise the temperature of the mating bodies and causes transformation of the surface material. The transformation product is also influenced by the environment around the components.

1.5. Scope of the Present Work

The present work focuses on investigating the effect of surface coatings/modification on fretting wear and fretting fatigue damage of Ti-6Al-4V and Ti-6Al-7Nb alloys. The surface modification processes used in this work are PVD TiN coating, plasma nitriding, nitrogen ion implantation, laser nitriding and thermal oxidation. Fretting wear is conducted with alumina ball and 1N normal force in air and Ringer fluid for all the coatings. Fretting fatigue is conducted with 40 MPa (600 N) contact pressure and axial stresses in the range of 67 to 500 MPa.

Chapter 2

LITERATURE REVIEW

2.1. PREFACE

The following section gives an extensive literature on biomaterials, titanium alloys used for biomedical applications, orthopedic devices, total hip arthroplasty (THA), fatigue and fretting fatigue of biomedical devices, fretting wear and fretting fatigue mechanism and its applications related to engineering materials including biomedical titanium alloys.

2.2. BIOMATERIALS

Biomaterials are a class of biologically compatible materials utilized for restoring the activity of damaged part of the human body, which can perform similar function of a natural system. They are widely used for repair, replacement or augmentation of diseased or damaged parts of musculoskeletal system. According to their period of service, they are classified as implants when they remain for substantial period of time and prostheses when they last until the end of lifetime. Orthopedic implants are normally fixed unto the skeletal system for restoring the lost function of the original part. Figure 2.1 shows some of the biomedical devices used in different parts of the body. It is also necessary for them to function without receiving any adverse tissue reactions or any inappropriate effects. Material suitable for implantation must therefore tolerate the aggressive environment and cyclic stress condition in case of load bearing joints. Orthopedic materials include ceramics, polymers and metallic alloys are commonly used for manufacturing medical devices.

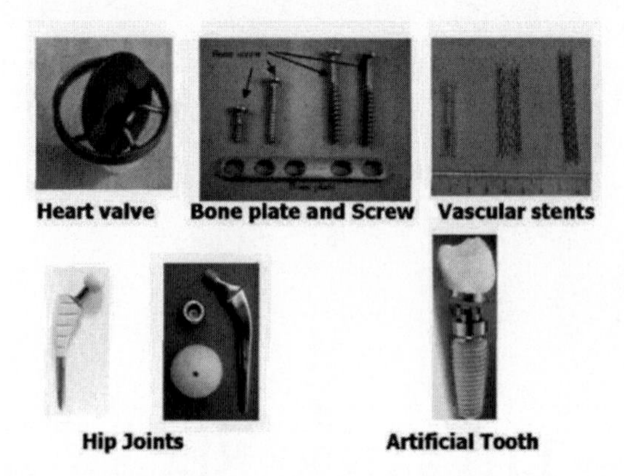

Figure 2.1. Biomedical devices for various applications.

Ceramics are inorganic compounds known for its good chemical stability within the body environment. They are carbon, alumina, zirconia, bioactive glass and calcium phosphate. The limitations of ceramic materials are their low tensile strength and fracture toughness. Polymers are normally used for hip sockets due to its close resemblance to natural tissue components. They are prone to degradation from biochemical and mechanical factors within the body environment resulting in formation of hydroxyl ion finally leading to allergic reactions. Fiber reinforced polymers have also been investigated for femoral components of hip joint arthroplasty and bone cements. They also suffer from degradation under complex states of stresses and low mechanical strength. Metals and alloys have wide range of applications ranging from dental amalgams to total joint replacements. The high strength and ductility of metallic systems makes them more suitable for load bearing implant applications due to reduced susceptibility for permanent dimensional changes in response to cyclic stress conditions. Three common alloys used for orthopedic applications are stainless steel (316 LV), cobalt-chromium alloy and titanium alloys. Amongst all the materials, titanium and its alloys are most commonly used for orthopedic devices (Marc Long et. al., 1998).

2.3. Titanium Alloys for Orthopedic Applications

Titanium alloys are attractive material in the field of orthopedic applications due to its lower modulus, excellent biocompatibility and corrosion resistance compared to other conventional materials previously used such as stainless steels and cobalt based alloys. The composition must be carefully selected to obtain the required properties as well as to avoid any adverse reaction from the release of elements. The metallic alloys used for fabrication of artificial joints undergo corrosion and release metallic ions into the patient's body. High corrosion rates and solubility of the corrosion products by release of metal ions by implants induces detrimental allergic effects. Elements such as V, Ni and Co are found to be highly toxic (Sumita et. al., 1994). Therefore, stainless steels and cobalt-chromium alloys are not suitable as implants for this reason. Whereas Ta, Nb and Zr are good biocompatible substitutes for these elements. Even Mo, Al and Sn are reported as not fully biocompatible elements (Yoshimitsu et. al., 2005, Marc Long et. al., 1998, Nadim et. al.,, 2003). The most often observed form of this allergy is skin rashes. The frequency of skin sensitivity to metals in patients with artificial joints is substantially higher than that in the general population (Hallab, 2001). The **β titanium** alloys with lower modulus of elasticity falls into the category of second generation alloys having made from more biocompatible elements such as Ta, Zr and Nb. Geetha et. al., 2003 gives a detail review on corrosion and microstructural aspects of titanium and its alloys. Some of advanced alloys are being tried for vascular stents, catherter guide wires, orthodontic arch wires and cochlear implants. Table 2.1 (Mitsuo Niinomi, 1998) gives the mechanical properties of all the medical grade titanium alloys currently in wide application for medical prosthesis.

Table 2.1. Mechanical properties of biomedical titanium alloys (Mitsuo Niinomi, 1998).

Alloy	Tensile strength (MPa)	Yield strength (MPa)	Elongation (%)	RA (%)	Modulus (GPa)	Type of alloy
Pure Ti Grade 1	240	170	24	30	102.7	α
Pure Ti Grade 2	345	275	20	30	102.7	α
Pure Ti Grade 3	450	380	18	30	103.7	α
Pure Ti Grade 4	550	485	15	25	104.1	α
Ti-6Al-4V ELI (mill annealed)	860-965	795-875	10-15	25-47	101-110	α+β
Ti-6Al-4V (annealed)	895-930	825-869	6-10	20-25	110-114	α+β
Ti-6Al-7Nb	900-1050	880-950	8.1-15	25-45	114	α+β

Table 2.1. (Continued)

Alloy	Tensile strength (MPa)	Yield strength (MPa)	Elongation (%)	RA (%)	Modulus (GPa)	Type of alloy
Ti-5Al-2.5Sn	1020	895	15	35	112	α+β
Ti-5Al-1.5B	900-1080	820-930	15-17	36-45	110	α+β
Ti-15Sn-4Nb-2Ta-0.2Pa (annealed) (aged)	715-919	693 806	28 18	67 72	94 99	α+β
Ti-13Nb-13Zr (aged)	973-1037	836-908	10-16	27-53	79-84	Near β
TMZF (Ti-12Mo-6Zr-2Fe) (annealed)	1060-1100	1000-1060	18-22	64-73	74-85	β
Ti-15Mo (annealed)	874	544	21	82	78	β
Tiadyne 1610 (aged)	851	736	10	-	81	β
Ti-15Mo-5Zr-3Al (ST) (aged)	852 1060-4400	838 1000-1060	25 18-22	48 64-73	80	β
21 RX (annealed) (Ti-15Mo-2.8Nb-0.2Si)	979-999	945-987	16-18	60	83	β
Ti-35.3Nb-5.1Ta-7.1Zr	596.7	547.1	19	68	88	β
Ti-29Nb-13Ta-4.6Zr (aged)	911	864	13.2	-	80	β

2.4. Total Joint Replacement (TJR)

It is generally understood from many years of medical investigation and experience that human joints such as hip, knee or shoulder joints are highly prone to degeneration leading to acute pain and joint stiffness commonly termed as osteoarthritis. Osteoarthritis develops slowly over several years. The symptoms are mainly pain, swelling, and stiffening of the joint. Pain is usually worse after activity, such as walking.

People, who suffer from chronic hip pain or degenerative joint disease of the hip, often have their ailing hip joint replaced with an artificial one. These artificial joints (or hip implants) are made of alloys, such as titanium or stainless steel, and have long stems which penetrate deep into the femur canal (center of the thigh bone) to hold them in place.

2.5. TOTAL HIP ARTHROPLASTY (THA)

In a surgical operation known as Arthroplastic surgery or Primary Total Hip Replacement (PTHR), the surgeon initially anesthetizes the lower part of the body below from the hip, while the blood is continuously supplied from the upper region of the body. The hip area is incised and the surgeon cuts off the natural joint and later drills a hole down the femural canal, and inserts the implant firmly into the hole, sometimes using special cement (PMMA) or a positioning ring (Figure 2.2) to ensure that implant stays tightly fixed in place. The surgery normally takes several hours and complete healing takes several months. Hip prosthesis is made up of a shaft, ball and socket that fit together to form a joint similar to natural joints. The parts of the hip shaft will be explained clearly in the later part of the chapter.

Various types of high-grade surgical metallic alloys used in combination with special metal/plastic/ceramic weight-bearing surfaces are cemented into place, reproducing a smooth articulating joint. An X-Ray negative of the implanted hip prostheses is shown in Figure 2.2 (Sumita et. al., 1994).

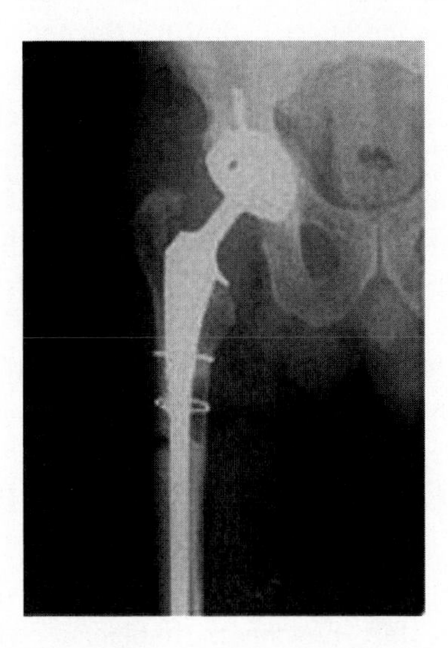

Figure 2.2. X-Ray of an implanted hip.

2.6. Fatigue Fracture of Total Hip Prosthesis

All the modern alloys used for manufacturing total hip prosthesis are strong enough to resist fatigue fracture from these repeating stresses in average.

There is a limit to how much repetitive loads the prosthesis will eventually sustain. This limit is specific for every form of the total hip prosthesis and for the metal alloy used for manufacturing the prostheses. Above this limit, the prosthetic shaft will not sustain the fatigue fracture. Closer examination of these cases revealed that the fractures occurred in heavy patients, often after an accident. The examination of the broken shafts sometimes reveals defects on the surface such as scratches and pits (Teoh, 2000). Many manufactures have also developed bulky models of artificial joints for heavyweight patients. Evaluation of fatigue behavior is normally done through fracture mechanics approach and in-vitro fatigue test. Fracture mechanics approach involves investigation of crack growth rate of pre-cracked compact tension specimens in physiological medium. In-vitro test for evaluating of fatigue resistance in physiological medium is more useful due to conjoint effect of corrosion and fatigue.

2.7. Stress Shielding

The prosthetic shaft takes off a part of the stress that walking and other everyday activities put on the upper part of the thighbone holding the prosthesis. The stiffness of the bone (10 - 30 MPa) is almost 7 times less than stainless steels (>200 MPa) and Co-Cr-Mo (230 MPa) alloys. The shielded bone thus loses its strengthening substance and becomes weak due to improper transfer of load from the prostheses. The total hip joint may then fail due to improper anchorage in a weakened femural canal. The remedy is to use a prosthetic shaft manufactured from metal alloys with stiffness similar to bone. Second generation titanium alloys (β grade) have the lowest stiffness among all other orthopedic materials and therefore shafts of cementless total hips are often made from these alloys.

2.8. Fretting Fatigue of Hip Prostheses and Bone Plates

Figure 2.3a shows hip prostheses implanted within the femur canal experiencing compressive cyclic loads due to body weight. Figure 2.3b shows fracture fixation device or bone plate with screws. Fretting areas are clearly indicated in both the figures. Modular joints and stem-bone cement interface are the two possible areas of fretting failures in hip prostheses. Figure 2.4 shows the post implantation failure incidences due to fretting fatigue in hip joints and bone plates. Modular junctions involve male/female tapered junction interlocked together due to friction. Metal-metal or metal-ceramic combinations are widely used as stem and ball heads. The combinations can be same or different in metal on metal couples. Cobalt-chrome heads on titanium stems are commonly used with cobalt-chrome, titanium, alumina or zirconia heads.

Kamachi et. al., 2003 have reported a detailed survey of failures of stainless steel orthopedic implants due to corrosion, fatigue and wear. According to their assessment, 74% of implants failed at the tapered junction of the shaft due to fretting fatigue. Every fretting cycle compounds the damage, generates harmful metal debris and exposes the bare metal surface to the surrounding corrosive bio environment. It also causes clearance at the contact. Adverse toxic effects have been reported even with 10 ppb of micro-debris particles (Teoh, 2000). Some of the prostheses are provided with metal fixtures to prevent the stem from turning.

In titanium implants, which are considered highly biocompatible compared to other materials, blackening of surrounding tissues have been reported, which is called metallosis (Sumita et. al., 1994). This may arise due to severe fretting conditions such as large variation in contact pressure and cyclic loads.

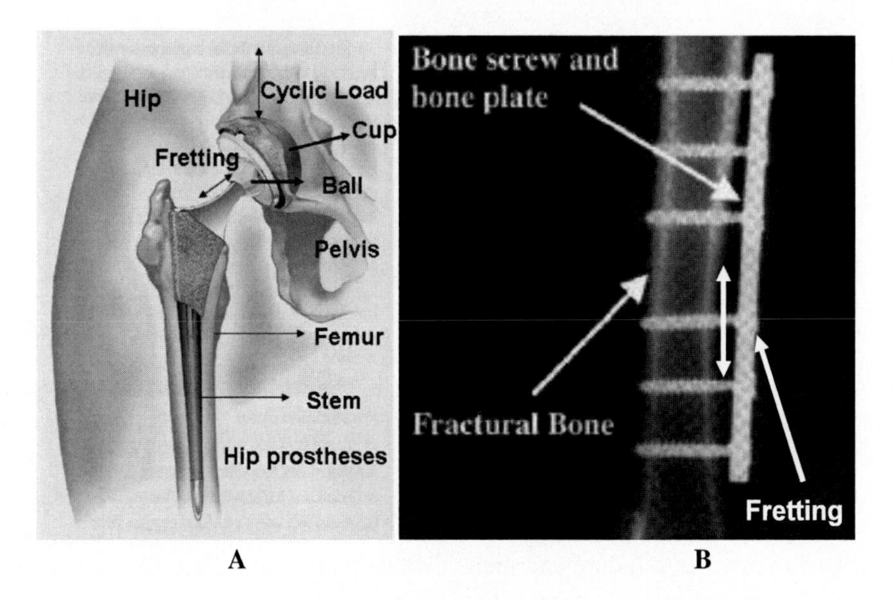

Figure 2.3. Fretting of (a) Hip implant, (b) Bone plate (Sumita et. al., 1994).

Therefore, either fretting action must be completely arrested or damage must be minimized by surface modification. The concentration of each metallic element is obtained from filtering the recovery liquid through certain membrane and analyzing the residue. Fretting motion experience lubricating conditions experiencing more dwell in the gross slip regime. Tangential force will be relatively reduced compared to fretting test in air unless the medium is highly corrosive causing the roughening of the surface. Longer dwell time may also aggravate the corrosion of exposed underlying metal.

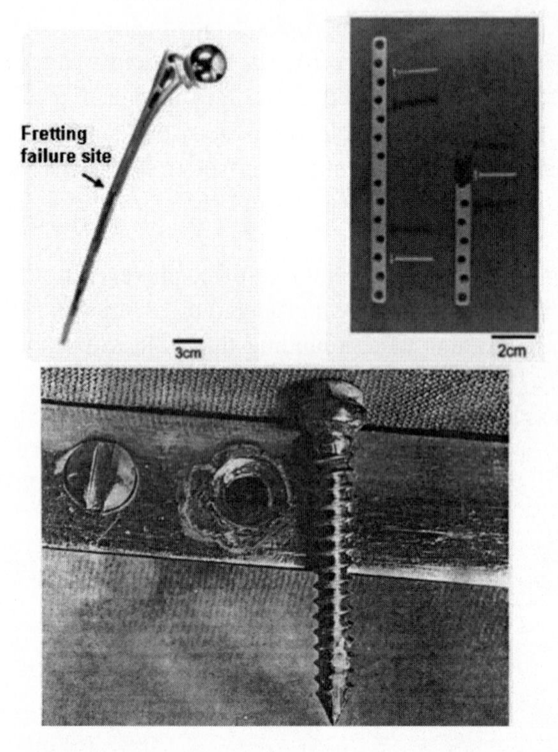

Figure 2.4. Fretting failure incidences of hip joints and bone plates (Sumita et. al., 1994).

From the electrochemical reaction point of view, fretting of surface would lead to fracture of oxide layers and drop in the OCP voltage (Hallab et. al., 2003). Its repassivation would increase the concentration of harmful chloride and hydrogen ions at the surrounding area due to consumption of oxygen. Both the mechanism would reduce the stability of the oxide film. Corrosion of implants will lead to premature failure and immediate tissue reaction. Most in-vivo studies indicate sustained release of metallic elements into the body fluids leading to infection and inflammation from toxic compounds. It is assumed that implant alloys continuously dissolves and reprecipitates on its surface. Repassivation time of the surface depends upon the material and surrounding medium.

Fretting fatigue failures can also be encountered by fretting damage of shaft from filler bone cement within the femur canal. Bone cement is filler material used for fixation of the artificial joints to the bone stock. It is a compound consisting of 90% polymethylmetacrylate (PMMA) and barium sulfate. The bone cement as prepared by the surgeon at the operation table is a material with many drawbacks. (a) It is mechanically weak because it has entrapped impurities such as air and blood, (b) It spawns small particles from its surface containing hard crystals of Barium sulfate, which scratch and damage the fine joint surfaces of the artificial joint. (c) It may also cause allergy and anaphylactic reaction during the operation. So fretting degradation is highly possible in this area.

Fretting studies of titanium alloys is more meaningful in a physiological medium representing the body fluids. Ringer solution, phosphate buffered solution (PBS), hank's balanced salt solution (HBSS) are some of the widely accepted mediums for preliminary biomaterial fretting fatigue studies. Starosvetsky et. al., 2001 have studied the corrosion behavior of 1 μm titanium nitrided coating on Ni-Ti alloy in ringer solution although the protein may influence the surface. Laure Duisabeau et. al., (2004) have conducted fretting wear test for Ti-6Al-4V and AISI 316L stainless steel in ringer solution. They remark that the presence of a solution containing chloride ions activates a localized corrosion phenomenon, which leads to modification of the displacement regime. Wear tests of biomedical alloys are normally conducted with dilute bovine serum in varying concentration. It is also observed that high temperatures generated during fretting will denature the protein and increases the viscosity of the medium. Although the fretting motion is affected by the proteins in the body, the effect of chloride ions is considered in these experiments.

Surface engineering of titanium alloys used for biomedical applications has a promising future from the view of minimizing in-service degradation of biomedical devices due to wear and corrosion. It also greatly improves the biocompatible nature of the materials to suit the surrounding body fluid environment. Some of the widely used processes of surface modification are described below.

2.9. Surface Engineering of Titanium Alloys

2.9.1. Nitriding of Titanium Alloys

Nitriding of titanium alloys has been investigated for many years. Nitrogen is found to significantly strengthen the surface layer due to its higher solubility in α-Ti. It has higher diffusion coefficients in titanium alloys. Introduction of nitrogen into titanium produces compound layers of TiN and Ti_2N with hardness of 3000 and 1500 VHN respectively (Ani Zhecheva et. al.,, 2004). Harder and biocompatible layers of nitrides are favorable for improving tribological properties of biomedical devices. The formation of compounds also depends upon the process being used to inject nitrogen into the material. It can be introduced by slow diffusion process aided by thermal energy or quickly through ion implantation.

2.9.2. Processes of Surface Modification

Many different processes are available for improving the surface properties of biomedical devices. They are selected based on their quality of improvement of implant surfaces. The following is a brief description of the processes normally used for surface modification of titanium alloys and the processes used for current experiments.

2.9.2.1. Physical Vapor Deposition (PVD)

PVD is a form of vapor deposition wherein a substance is evaporated, and then condensed onto a substrate for the purpose of coating the substrate. This process is generally performed in a vacuum. The basic idea behind physical vapor deposition (PVD) is the physical change of state of the material that is being deposited (target material). There are two main ideas that are used to obtain a vapor state of the material being deposited. There is evaporation of the material either through thermal heating of a filament or through melting the material with an electron beam or laser. The second method that is used is to physically remove the target material atoms with the transfer of momentum from an incident particle such as ions. This type of material removal is called sputtering. The coatings can be from a single element or a compound. Compounds can be produced by mixing the evaporated element with another element usually in the form of gas. This process is widely used for developing titanium nitride (TiN) coating on titanium alloys and has been under investigation since many years for biomedical applications.

Huang nan et. al., 1996 have studied the fatigue behavior of titanium based biomaterial coated with 1.4 μm titanium nitride film by ion beam enhanced deposition. Shenhar et. al., (2000) have characterized residual stresses and fretting wear behavior of 2 μm TiN coating developed on surgical titanium alloys by powder immersion reaction assisted coating method. A major reduction in fretting damage was reported from the coated alloys. In these experiments, an attempt is made to study the fretting wear and fretting fatigue behavior of 2 μm thick PVD TiN coated titanium alloys.

Zenghu Han et. al., 2004 have investigated the effect of coat thickness and substrate (high strength steel and stainless steel) on the mechanical properties of PVD TiN coatings using nano indentation technique. He concluded that hardness and modulus of TiN vary with different substrates and different coating thicknesses. TiN coatings on harder substrates or with higher thicknesses show greater hardness and elastic modulus. These effects can be attributed to the internal stress of TiN. TiN coatings are generally characterized by high abrasion resistance, low coefficient of friction, high temperature stability and high hardness. These properties also make it more biocompatible in the human body. It is observed that TiN with (111) preferred orientation has the highest hardness governed by thickness and residual stresses of the film (Wen-Jun Chou et. al., (2003)). The ultimate interfacial shear stresses between coating and substrate can be evaluated from simple tensile test and measurement of surface cracks (Wen-Jun Chou et. al., 2003, Takahito Ohmura et. al., 2003).

TiN has a NaCl type structure with the active slip system of {110} <110> (Hultman et. al., 1994). Plastic deformation is possible if resolved forces act on this plane. If the external forces are perpendicular to (111) system, then it is difficult for the deformation and hence higher hardness are obtained. At higher temperatures (> 200 °C), the hardness of the substrate decreases and the wear rate rapidly increases (Badisch et. al., 2003). Post heat treatment of coated samples also sometimes improves the hardness of the coating due to structural rearrangement (Wen-Jun Chou et. al., 2003).

2.9.2.2. Plasma Nitriding (PN)

It is a thermochemical treatment by which nitrogen is ionized in vacuum and allowed to diffuse into the substrate maintained at high negative voltage. Plasma nitriding (also known as glow discharge or Ion nitriding) is a low temperature; low distortion surface engineering process. Glow discharge plasma is used to transfer nitrogen ion into the surface of the components undergoing treatment. It requires specialized equipments with high ionizing energies. The components are made the cathode of an electrical circuit whilst the chamber becomes the anode; application of a voltage between the 2 electrodes establishes a current intensive glow discharge. This anomalous glow discharge covers the entire cathode supplying heat to the surface of the parts and a supply of nitrogen. Nitrogen is able to diffuse into the surface where it combines with nitride forming elements to form alloy nitrides. Bekir et. al., 1996 has investigated plasma nitriding of Ti-6Al-4V and found a dramatic improvement in wear and frictional properties from the presence of δ-TiN and ε-Ti_2N.

2.9.2.3. Laser Nitriding (LN)

Laser beam nitriding is also a specialized technique meant for improving the surface of materials apart from cutting, melting and welding applications. Laser nitriding works by melting the surface using a focused beam in nitrogen atmosphere to form hard nitride layers. Nitrogen is fed through nozzle held at 30 degrees into the melt pool. The process parameters such as laser pulse energy, scanning speed and gas flow rate and concentration determines the thickness of surface layer compounds. It requires special equipments and also dependant on the geometry of the process. Man et. al., 2005 have studied laser nitriding on Ti-6A-4V and reported the formation of three dimensional networks of TiN dendrites.

2.9.2.4. Ion Implantation (II)

Ion implantation is a high technology approach for modifying surface properties of materials. Unlike coating process, it does not involve the development of layer on the surface. It was initially developed for use in semiconductor applications and in fact still used extensively in that capacity today. Ion implantation utilizes highly energetic beams of positively charged ions to modify surface structure and chemistry of materials at low temperature (< 180°C). The process does not adversely affect component dimensions or bulk material properties. Early applications of ion beams were removal of material (now called etching) and deposition using non-reactive beams. From the 1980s until today, their use have eventually progressed to reactive processes and property modification. Surface properties improved with Ion Implantation are wear resistance, corrosion resistance, hardness, reduced friction, fatigue and fretting fatigue resistance, oxidation, adhesion, and decorative finish (Elder et. al., 1989). The process can be applied to virtually any material, including most metals, ceramics and polymers. However, the effects of the process are typically material-specific. Examples of components treated with ion implantation are Ti and Co-Cr orthopedic prostheses, which are made harder and more wear resistant with the process and silicone rubber catheters, which are made less tacky and more water wettable for improved insertion and biological compatibility. Currently the high cost of process has restricted their use in medical implants, nuclear fuels, jet engine parts and expensive tools.

The three hardening mechanisms of ion implanted titanium alloys are: ion induced damage, solid solution hardening, and precipitation hardening. Ion induced lattice damage creates dense network of dislocations similar to cold worked structure. The strain fields make the deformation of the lattice network more difficult thereby improving tribological properties. In solid solution strengthening, the same strain fields are obtained through interstitial or substitutional impurity atoms. In precipitation strengthening, the dislocation movements are hindered by coherent or incoherent precipitates. Precipitates are usually observed for fluence greater than 10^{17} ions/cm^2. Solid solution hardening is achieved by fluence in the range of 10^{16} – 10^{17} ions/cm^2. The hardening achieved by surface compressive residual stresses or lattice damage is obtained with the fluence less then 10^{15} – 10^{16} ions/cm^2.

2.9.2.5. Thermal Oxidation (TO)

TO is a novel, cost effective and environmentally friendly surface engineering technique for titanium or titanium alloys. It can be conducted in ambient air. It is thermally aided diffusion of oxygen at elevated temperatures. Several investigations have been done on characterization of TO layer in Ti-6Al-4V alloy (Hasan Guleryuz et. al., 2004). TO can provide a titanium surface with dramatically reduced coefficient of friction and galling tendency when in rubbing contact with another engineering material surface, especially under lubricated conditions. Furthermore, the load bearing capacity of the surface is greatly increased. TO process can not only significantly improve the tribological behavior but also the corrosion resistance of titanium and titanium alloys. The surface is chemically inert and biologically compatible with cells and tissues and is ideally suited to use in surgical implants.

The oxygen atoms that diffuse into the metal can occupy both interstitial and substitutional sites, making the material harder, and since the concentration of oxygen at the surface would be much greater than interior, it should be harder near the surface and relatively softer inside. The correlation between the concentration profiles and hardness profiles suggests that microhardness testing can be an effective tool in studying the oxygen diffusivity in titanium.

2.10. Studies on Specific Titanium Alloys

2.10.1. Ti-6Al-4V (ASTM 1472)

It is one of the most popular α+β titanium alloys developed in the early 1950s at Illinois Institute of Technology, USA. It has good balance of mechanical properties and by far the most intensively developed and tested titanium alloys. Aerospace industries are the largest users of these all alloys. It is also very well suitable for medical applications compared to stainless steels and cobalt-chrome-moly alloys due to low stiffness and excellent biocompatibility, two most important factors for designing medical implants. Although they possess excellent biocompatibility and corrosion resistance, the presence of Al and V are known to cause cytotoxic effects and neurological disorders in its long term applications as virgin implants. Conjoint action of corrosion and wear is studied through fretting corrosion potential measurements of real implants in a simulated body solution. The various oxide films on the surface may also play a significant influence on adhesive wear of mating couple. Surface modified alloys have served far better compared to virgin alloys due to its superior resistance to wear and corrosion. The case depth, hardness and uniformity of the modified layer determine the capacity to resist wear damage. The implants such as hip joints or knee joints have to function for longer periods in younger patients. Surface modified Ti-6Al-4V promise a great future in coming generation of new medical devices.

2.10.2. Ti-6Al-7Nb (ASTM 1295)

This α+β alloy is especially designed for medical implants. Aluminum stabilizes α phase. Niobium is a biocompatible element and stabilizes β phase as discussed earlier. The wear resistance of these alloys is also poor and needs enhancement of surface properties. The biocompatibility of the surface also depends upon the nature of oxides present. Al_2O_3 and Nb_2O_5 are normally found on the surface of these alloys (Sittig et. al., 1999). Ion implantation and plasma nitriding on these alloys has been investigated by several researchers who established the superior quality of the surface in terms of wear and corrosion resistance due to formation of nitrides (Thair et. al., 2002). Alloys for load bearing applications need special consideration in terms of wear and fatigue. Surface modification must not induce defects in the highly stressed parts of the implants. Hard and brittle nitride compounds are prone to initiate failures if the loads exceed critical value.

2.11. DETAILS OF FRETTING WEAR AND FRETTING FATIGUE MECHANISMS

Wear mechanism are classified into adhesive wear, abrasive wear, three body wear, fatigue wear and corrosive wear. Two mating bodies exhibit adhesion when the atomic forces occurring between the materials in two surfaces under relative load are stronger than inherent material properties of either surface. It is related to repeated formation and breaking of bond junction between contacting asperities. Fretting is a form of adhesive wear between contacting bodies under normal load and relative motion. Fretting wear tests are normally conducted for the purpose of ranking the wear resistant nature of materials or to understand the underlying damage mechanisms of the wear phenomena. It may also serve the purpose of understanding the wear damage characteristics of alumina and titanium couplings used for hip implants, intramedullary nails etc. The following discussion on fretting wear may serve all the purpose. Fretting fatigue tests is conducted to study the damage mechanism of surface modified titanium alloy couplings. The following section gives some literature details about fretting wear and fretting fatigue failures.

2.11.1. Fretting Wear

Fretting wear damage is predominant in many areas of engineering applications such as power generation industries, electrical contacts as well as biomedical applications. Young Ho Lee et. al, 2005 have studied the fretting of Korean power plant steam generator tubes, which is made of Inconel 690 coupled with ferritic and martensitic steels as supports. Some of the grade of Inconel alloys could resist plastic deformation much better compared to others. SEM observation revealed that fretting condition induced extensive plastic deformation at the local area and generated debris in the form of thin plates and particles causing third body wear mechanism. Microstructural constituents also influence the fretting wear damage mechanism. Jin-Ki Hong et. al., 2005 have studied the effect of microstructure on fretting wear of Inconel 690. Larger grain sizes (60 to 80μm) with coarser carbides along the grain boundaries have good resistance to fretting wear. However, they are also held responsible for formation and propagation of cracks. Sung et. al, 2001 studies the effect of TiN coating in reducing fretting wear damage of Zircaloy-4 tube. It is reported that the main wear damage mechanism of TiN coating was the brittle rupture of the film followed by low slip amplitude fretting and cumulative plastic flow at the contact edge. Adhesive and abrasive damage followed by oxidation was prominent at high slip amplitude.

In the area of biomedical applications, Animesh et. al, 2004 have investigated the tribological behavior of three different titanium based alloys which indicates that the major wear mechanism of titanium alloys is tribomechanical abrasion, transfer layer formation and cracking due to extensive plastic deformation under tribomechanical stress conditions and persistent slip band formation. Chen et. al, 2002 have investigated the fretting wear characteristics of CrN and PVD TiN coating coupled with corundum ball at different relative humidities. It is reported that TiN coating exhibits high friction in dry conditions i.e., at reduced relative humidities compared to high humidities. Contact friction is also very much related to the type of debris. Chen et. al, 2005 indicates that low coefficient of friction coincides with nanocrystalline debris and high coefficient of friction with amorphous debris. Anatase or rutile is commonly formed as debris during fretting of PVD TiN coatings. Tan et al, 2005 have studied the effect of oxygen implantation on wear behavior of NiTi shape memory alloys indicating improvement in wear resistance of the alloy.

Ion implanted materials also shows remarkable improvements in fretting wear resistance with reduced friction (Tan et. al., 2005, Prakash et. al., 2003). The case depth of implanted materials is shallow and easily worn off within few cycles unlike thicker coatings. Improved adhesion of TiN layer can also be obtained when ion implantation is coupled with PVD process (Gunzel et. al., 2001).

Plasma coatings are reported to improve wear, corrosion and fatigue strength of biomedical implants (Hong et. al., 2004, Chu et. al, 2002, Sobiecki et. al, 2001). Plasma nitrided surface gives superior surface textures with no significant geometric distortions (Alves et. al, 2005). The wear debris formed during fretting is predominantly titanium oxides and titanium oxynitrides (Gokul Lakshmi et. al, 2004). The temperature plays more important role than other parameters like treatment time, gas flow ratio etc.(Shengli et. al, 2004, Mishra et. al, 2003). The material removal in saline medium for plasma nitrided titanium alloys in is influenced by both mechanical wear and electrochemical oxidation (Galliano et. al, 2001). More than 50% of the wear volume generated is attributed to electrochemical oxidation during fretting process.

Laser nitriding of titanium alloys involves melting and introduction of nitrogen species within a few micron layers from the surface before resolidification. The hardness profile of the modified layer is very much dependent upon nitrogen concentration gradient and interaction time (Carpene et. al., 2005). Laser nitriding generates 3D network of TiN dendrites (Man et. al.,

2005, Mohamad et. al., 2001). The cracking tendency is influenced by the volume fraction of TiN dendrites and preheating the specimens before nitriding can reduce such tendencies and also bring about uniformity in case depth (Hu et al, 1999). Tribological properties of laser nitrided alloys are better compared to unmodified alloys (Ettaqi et. al., 1998). The resistance is mainly offered by the TiN dendrites formed during the process.

Thermal oxidation involves gradual diffusion of oxygen into titanium substrate material **producing hard α titanium oxide** layer. Even water vapors can be utilized as a source to produce surface titanium oxides (Yves et. al., 1997). The wear, corrosion and galling resistance of titanium alloys can be greatly improved with thermal oxidation process due to reduced tendency for adhesion (Borgioli et. al., 2005, Guleryuz et. al., 2005, Güleryüz et. al., 2004, Dong et al, 2000, Wiklund et. al., 2001). The chemical composition of the oxidized layer depends on the temperature and thickness of the film. Below 200°C the oxidation process produces predominantly TiO_2 and at higher temperatures the presence of Ti_2O_3 was detected from the beginning of the process (Vaquila et. al.,, 2001).

2.11.2. Mechanism of Fretting Wear Damage

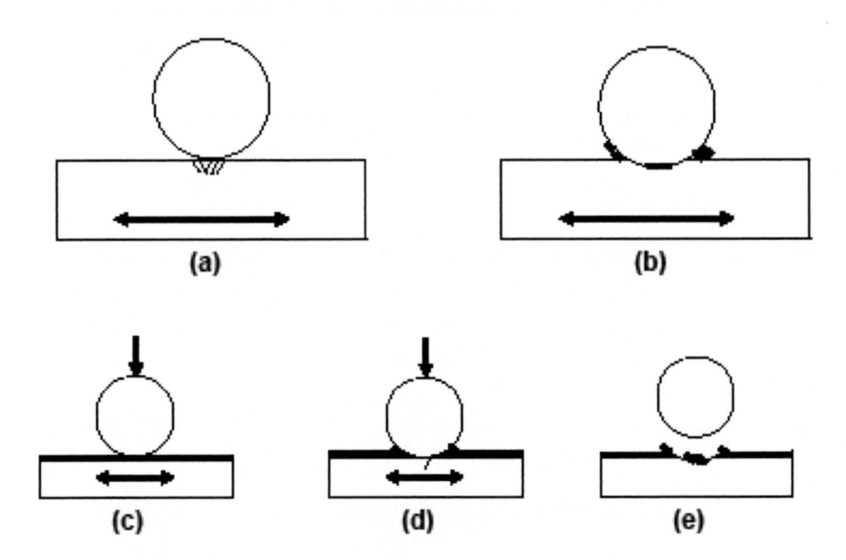

Figure 2.5. Fretting wear damage mechanism.

Figure 2.5 shows the possible mechanism of damage occurring during fretting wear of unmodified and surface modified alloys. In unmodified condition, fretting motion generates wear particulates according to the mechanism shown in Figure 2.5a.

Multiple cracks starts within the contact area inclined to the surface and later delaminate into chips and flakes of different sizes once these cracks turn parallel to the surface due to tangential force generated during fretting.

The generation and interaction of multiple dislocations within the subsurface layers may be responsible for propagation of fine cracks to form debris. Two body wear change to three body wear mode with the particulates in the contact. Continued sliding would further squash and grind them within the contact which is compacted within the cracks or ejected out of the contact area as shown in Figure 2.5b. Finer particles would transform to different oxide products during mechanical attrition. Wear products harder than parent material can also augment the severity of damage. Wear volume depends upon the loading and contact conditions. The generation as well as structure of wear products also depends upon the medium of fretting. The damage is dependent on physical nature of the debris so formed.

In case of surface modified materials, fretting motion initially damages the layer until it reaches the base material as shown in Figure 2.5d. The damage is delayed until the layer is completely worn off. Thicker and harder layers can offer resistance for a long period. The debris generated from the modified layer may or may not influence the fretting behavior. Fretting wear tests with ball contact configuration generates fine crater on the surface after the test as shown in Figure 2.5e.

Electrochemical activity can also influence fretting behavior depending on the medium used. In medical applications, fretting of implants takes place within the body in presence of saline and protein medium. Chloride corrosion is always a problem with some of the medical devices used in the body. Fretting normally enhances the corrosive activity by exposing fresh surface to the surrounding medium.

2.12. Fretting Fatigue Process

The following section gives some details about the fretting fatigue mechanism, fretting maps and the variable that affect the mechanism.

2.12.1. Mechanism of Fretting Fatigue Damage

Plain fatigue and fretting fatigue life is compared and presented in the form of Stress-Number of cycles (S-N) curves. Fretting fatigue lives are normally lower than plain fatigue lives due to premature damage caused by the contacting body as shown in Figure 2.6a. The total fretting fatigue life is divided into fretting and fatigue cycles. Both of them are again influenced by different set of parameters. Fretting cracks usually initiate within few cycles of operation and further growth of these cracks is controlled by axial fatigue load cycles.

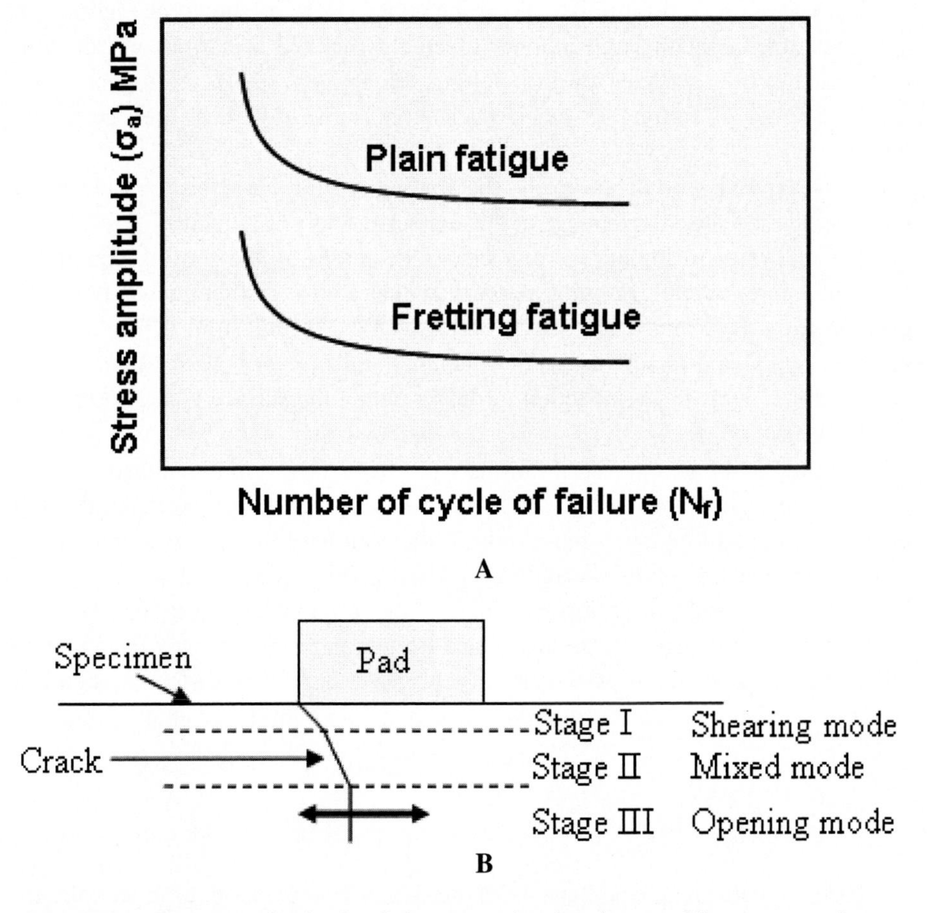

Figure 2.6. (a) S-N curve for fretting fatigue tests showing decreased lives compared to plain fatigue lives. (b) Modes of crack initiation and propagation during fretting fatigue test (Endo et. al., 1976).

In all the cases, initially it is observed that these cracks grow inclined to the tensile axis from the surface of the material and then propagate normal to the tensile load when the effect of tangential force diminishes when the crack reaches the bulk leading to ultimate fracture as shown in Figure 2.6b (Endo et. al.,, 1976). The three modes of crack propagation are also shown in the same Figure 2.6b. For high strength materials such as TiAl, the fatigue lifetimes were suggested to be determined by the cycles needed for initiation of crack rather than propagation (Hansson et. al., 1999). Here the SN curve is flat in contrast to metallic materials.

The growth rate of cracks is governed by microstructure of the base material, and bulk fatigue loads.

They are observed to initiate along the contact edge as shown in Figure 2.6b and sometimes contain large number of cracks inclined at various angles with compacted debris particles during fretting (Hoeppner, 1994). At higher stress levels the role of fretting in crack nucleation is overridden by persistent slip bands.

However at lower stress levels, the fretting motion has greater influence in reducing life, because the damage at the contact acts as crack nucleation sites. The number of cycles to initiate fretting cracks decreases as the fretting conditions becomes more severe. Fretting process will lead to particle generation and ejection from the contact. Fretting particles would generate from deformation of surface layers and crack assisted delamination. In some cases microscopic observation reveals the compaction of debris within the surface cracks (Antoniou et. al., 1997).

Fretting action can produce various types of surface symptoms distinct from ordinary wear phenomenon by which it can be diagnosed. For example, the rust debris on steel will be much more reddish than ordinary rust or aluminum debris is black instead of white (Waterhouse, 1972). This indicates that the fretting products may result in transformation of the original structure or chemical composition of the parent materials. Since fretting damage is mostly encountered under heavy pressure, it will certainly produce debris of different chemical structure depending upon the environment around the joint (Chan et. al., 2002).

2.12.2. Fretting Maps

Fretting damage phenomena of a particular combination of materials are normally quantified by what is called fretting maps. Vincent et. al., 1992 have given running condition and material response maps for fretting fatigue tests. Running condition fretting maps (RCFM) as shown in Figure 2.7a demarcates stick, partial slip and gross slip regions for a particular contact material and number of cycles. Material response fretting maps (MRFM) demarcates no crack, crack and debris formation regions for the same material as shown in Figure 2.7b. Frequency and roughness also influence the regime of fretting operation.

Gross sliding regime corresponds to dissipative process (large area of Force-Displacement (FD) curve) where the members slide without any obstruction. Area within the FD curves indicates the energy lost by the system. Gross slip is normally characterized by low contact pressure and longer strokes. The sliding tangential force is small enough not to cause any destruction of surface films. For larger slip amplitudes, wear phenomena predominates over fretting. For brittle materials such as ceramics, it normally corresponds to one cycle.

Partial slip or mixed stick-slip region is the most dangerous from the crack initiation point of view due to destruction of surface films and interface adhesion. Tangential force increases due to such adhesion of underlying exposed metal surfaces. Relative motion is frequently checked by micro-welding and rupture of asperity junctions indicated sometimes by squeaking and squealing sound during operation. It is commonly observed during fretting of metallurgically compatible pairs. In combination of the above phenomenon, it is difficult to predict the fretting fatigue lives of the materials. Fracture mechanics approach sometime helps us to understand the process better (Mutoh et. al., 1989). Stick regime is encountered at higher contact pressures as seen from Figure 2.7a. It is characterized by non dissipative elastic accommodation of fretting movement.

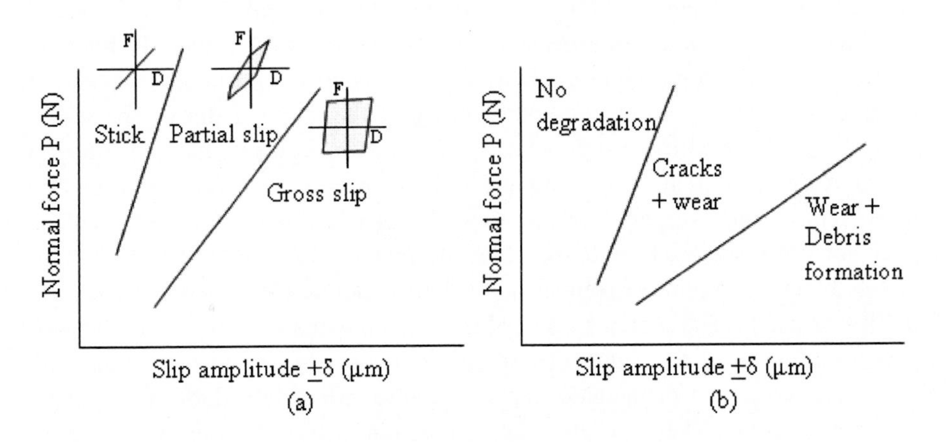

Figure 2.7. Fretting maps (a) Running condition fretting maps (RCFM), (b) Material response fretting maps (MRFM).

Due to reduction of sliding motion, the fretting fatigue life is sometimes increased in this region.

2.12.3. The Effect of Variables on Fretting Process

The following section gives some of the most important variables that affect both fretting wear and fretting fatigue mechanisms. It is a general description and hence applicable to both titanium alloys as well as other materials. All the information contained in the following section pertains both to biomedical area as well as to other engineering applications due to limited availability of literatures in the area of biotribology.

2.12.3.1. Fretting Media

The process of fretting can often be aggravated by the medium of its action depending upon the load and frequency of operation. The nature and volume of debris produced depends upon the aggressiveness of the environment. Oxidation or corrosion of materials can influence the fretting wear. Fretting action alone may provide sufficient driving energy for the atoms on the contact surfaces for the chemical reaction even at low temperatures. Fretting in vacuum or inert atmosphere produces much less damage due to absence of chemical reaction that weakens the surface strength. Fretting of steels in oxygen or air indicates more damage than nitrogen or helium atmosphere (Waterhouse, 1972). Dissolved oxygen in these fluids significantly influences the test results. It must be controlled strictly to represent the body fluid as genuinely as possible. N_2 and O_2 mixture is normally introduced continuously in the solution during the test to control the dissolved oxygen content. Fretting of bio implants made of titanium alloys is of very great concern in the area of orthopedic surgery due to high manufacturing and surgery cost. Fretting debris generated from these implants elicit adverse tissue reaction. It also depends upon the corrosion rate and solubility of the corrosion products. Human body fluid contains blood, tissue fluids, and 0.9% amino acids and proteins. The pH is 7 and body temperature is 37°C at 1atm pressure (Sumita et. al., 1994). Lower partial pressure of oxygen within the body accelerates corrosion of metallic implants due to reduced availability of oxygen for repassivation. There is also an assumption that corrosion is caused by increasing concentration of H^+ and Cl^- ions in the solution. Therefore bioimplants within the body must withstand the corrosive action of these chemicals along with fretting. This is another broad area of study under the name fretting corrosion. Khan et. al., 1999 comments that the wear of titanium alloys increase in the presence of proteins. The study on conjoint action of wear and corrosion on titanium alloys indicates that Ti-13Nb-13Zr alloys perform better than other alloys. Resistance to corrosion action can also be improved by some other specialized processes such as ion implantation (Thair et. al., 2002) and powder immersion reaction assisted coating (PIRAC) (Shenhar et. al., 2000). PIRAC nitriding reduced fretting induced metal dissolution up to 25%.

2.12.3.2. Frequency

Frequency of vibration due to dynamic loading is an important factor in structures that undergo fretting damage. It is reported that damage of mild steel specimens in weight loss are severe at lower frequencies (< 1000 Hz) due to magnified activity of chemical forces acting on the exposed fretted surfaces for longer time (Waterhouse, 1972). Where as, frequency has no effect in non reactive atmosphere (vacuum tests). Papakyriacou et. al., 2001, from their investigation on HCF studies of annealed and cold worked niobium and tantalum and also commercially pure titanium and Ti-6Al-7Nb alloy, concludes that there is no statistically significant influence of cyclic frequency on the endurance data and the mean fatigue limit. However, the fracture initiation mode changed from ductile transgranular mode at lower frequency to brittle intergranular mode at higher frequency. Rebecca et. al., 1999 indicated that constant amplitude higher frequency fatigue conditions on Ti-6Al-4V alloy produced shorter fatigue lives then those conducted under low-frequency fretting fatigue conditions. The effect of frequency on bioimplants within the simulated body fluid also depends on the severity of media surrounding the implant surface. Test frequency of 1 to 5 Hz is normally selected for bio implant fretting fatigue test (Hallab, 2003). There is a possibility that corrosion rate due to chemical reaction of exposed metal surfaces during low frequency fretting may be higher than the higher frequencies due to longer dwell period of exposed damaged area.

2.12.3.3. Slip Amplitude

Relative slip amplitude measured between the two fretting members is an important parameter because the volume of material removed increases linearly with the amplitude of slip. In some cases, wear loss increases rapidly above certain limit of slip amplitude and the damage is very severe. The slip amplitude depends upon the range of alternating stresses and strains in the fatigue loaded specimen. Variation of slip amplitude with friction coefficient is a useful plot from the design point of view. Vincent et. al.,, 1992 gives running condition and materials response fretting maps for 2091 aluminum alloys clearly demarcating the three regions viz. stick, partial slip and gross slip regions. It is observed that all materials are covered with some form of protective layer that comes in contact with each other during initial stages of fretting. Fretting life is very much influenced by the properties of such layers. Such layers can also be produced using surface engineering techniques to improve the fretting life of materials which will be discussed later. During first few cycles, the friction is low and gross sliding is prominent. In this stage the natural films rub each other until they break away from the surface after certain period and the force displacement plot assumes trapezoidal form. After several cycles, the natural film rupture and the bare metals underneath are exposed to each other. In this partial slip stage friction increases due to adhesion of two bare surfaces. Elastic accommodation of slip is increased here resulting in damage of the surface films. Surface modification with stronger films will increase the period of rupture of the protective coatings thereby delaying the shift of gross slip regime. Apart from the surface films the shift in the fretting regime is also dependent upon the surrounding media. With regard to bio implant, the shift in the regime may be much more delayed due to lubrication effect of the protein serum or may also aggravate due to corrosive nature of the fluid. Therefore the contact behavior is primarily governed by particle detachment and debris ejection from contact area. The initial surface roughness also plays a significant role in controlling slip amplitude during fretting process.

2.12.3.4. Normal Pressure

Normal pressure is the stress applied by the pads perpendicular to the surface of the fatigue specimen during fretting process. The magnitude of normal pressure affects the shift in the region of fretting maps. However the effect of normal pressure is always felt depending on the material and surface condition of the contacts. If contact pressure is high enough, it may cause premature rupture of surface film due to high friction and expose the underneath bare metal earlier causing adhesion.

Figure 2.8 shows the effect of normal pressure on fretting fatigue life of Ti-6Al-4V (Kozo Nakazawa et. al., 2003). Three normal pressures are selected for 1.65kN maximum cyclic load and one normal pressure is selected for 3.30kN maximum cyclic load. It is evident from Figure 2.8 that increase in normal pressure has reduced the life of hip prostheses under cyclic loading conditions. Higher cyclic load (3.30kN/0.330kN) induces more severe fretting due to increase in relative slip amplitude between the contact pairs.

This has drastically reduced the life of the material for the same contact pressure (41.4 MPa). Hager et. al., 2004 have characterized the mixed and gross slip wear regimes in Ti-6Al-4V to identify the critical normal load and displacement corresponding to these regimes. Their experiments have shown that there is a linear relation between the load and displacement. Galling and cold welding were much evident on the fretted surface at the critical values. At modular joints, it is expected that load would reach two to three times the total body weight and extreme pressures are generated along the contact during fretting action (Kamachi Mudali et al, 2003). It may also vary because the hip joints are available in wide variety of designs and modular area varies for each design. Some of the hip prostheses come as a single piece with the entire model made of single alloys. There are no modular junctions and fretting problems are expected at the stem inside the femur canal where the bone cement or positioning ring is in contact with the shaft.

Even the compacted debris within the contact influences the variation in pressure throughout the test. The contact pressure varies according to the nature of the transformation product. Rutile formation is unavoidable in case of fretting of titanium alloys unless it is perfectly insulated from the ingress of oxygen.

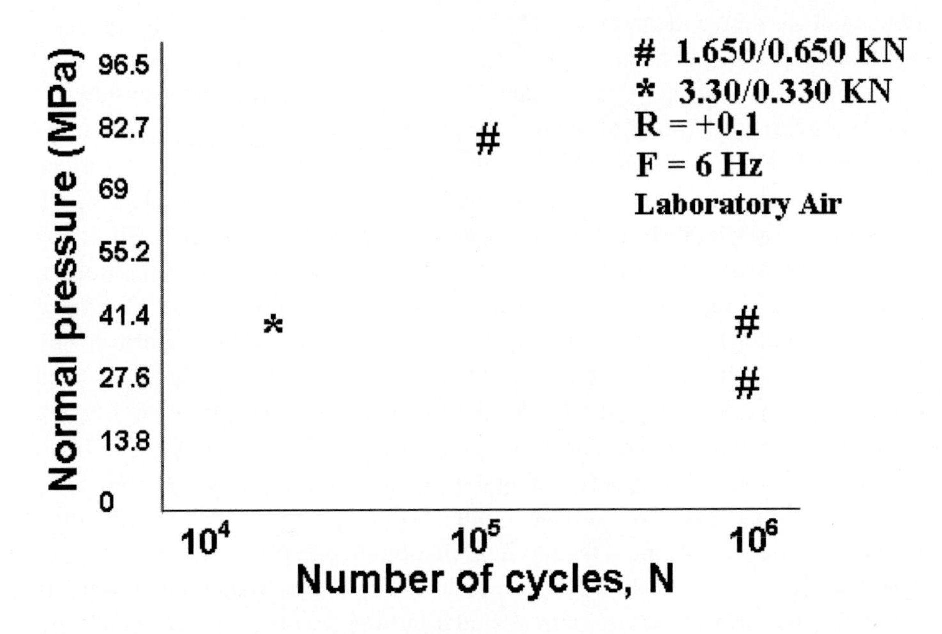

Figure 2.8. Ti-6Al-4V fretting fatigue test with hip prostheses loading conditions, (Kozo Nakazawa et. al., 2003).

In the case of alloys made of similar materials the friction is severe because the contact pairs are metallurgically compatible. The specimen damage is considered more importance because the effect of axial load is more prominent in the specimen than pads which constitute failure.

2.12.3.5. Hardness

With regard to hardness, there should be a balance between increase in the fretting life due to higher hardness of the surface and decrease in the life due to decreased toughness lowering the energy required to initiate fretting fatigue cracks. High surface hardness can effectively prevent fretting wear damage, whereas it can cause premature failure of cyclically loaded member during fretting fatigue. Therefore, there should be a balance between the hardness and thickness of the modified layer.

2.12.3.6. Surface Finish

Surfaces finish play a vital role in the fretting process as described in many sections earlier. Normally even very smooth surface will have defects at the atomic level. The defects are normally sharp projections and asperities of different contours. This causes the unevenness of the surfaces and it is practically difficult to achieve crystallographically flat surfaces in metals and alloys even with high degree of surface finish. The real area of contact between mating surfaces is considerably reduced due to higher roughness and the wear debris can be accommodated with in the asperities. As mentioned earlier the decrease in surface roughness results in increased friction and increase in surface roughness will produce asperity interlocking resulting in decreasing fretting amplitude. Since this process is non-dissipative, elastic accommodation relives the surface from fretting process. The hip implant ball is normally completed to a high degree of surface finish to minimize the wear debris generation. It is further improved by specialized surface modification techniques to improve the hardness and biocompatibility. This will reduce frictional forces between surfaces. For the industrial components lubrication coatings such as MoS_2, DLC or CuNiIn films are used to reduce friction (Yongqing Fu et. al., 1998). Bulatov et. al., 1997 produced fretting resistive surface from “Vibro-rolling” method containing regular roughness pattern. Vibro-rolling involves fine plastic deformation of subsurface layers with hard tool (hardened ball or rounded diamond tip) which is forced into the surface and passed in a regular fashion creating different roughness pattern.

2.12.3.7. Contact Materials and Microstructure

Plain fatigue life will be superior for fine grain than coarse grained material. For c.p. titanium, the endurance limit is increased by 70 MPa when the grain size is reduced 20 times (Leynes and Peters, 2004). The surface microstructure plays an important role in initiating the fretting induced cracks. A reduction in the grain size increases plain fatigue strength and coarse grains reduce the fatigue crack growth rate. Even oxygen content bears influence on crack growth rate. Hyukjae et. al., 2004 investigated the effect of dissimilar pad materials (Aluminum alloy 2024 and Inconel 718) on the fretting fatigue behavior of Ti6Al4V. At lower contact forces, there was no effect on fretting damage. When the contact forces increases, the hardness effects overrode the effect of slip amplitude and coefficient of friction of the mating material. Therefore relative hardness of the contact materials makes the difference. Hoeppner et. al., 2003 also investigated the effect of Al-1100, Cu, 0.40/0.50 steel and Ti-6Al-4V pads on Ti-6Al-4V alloy. Significant reduction in fatigue life was obtained by Ti-6Al-4V in contact with Ti-6Al-4V. Sinha et. al., 2001 investigated the effects of colony microstructure on fatigue crack growth rate of Ti-6Al-4V. The growth rate reduces with increase in the alpha lathe size and colony size. So fretting fatigue life in such a case would also improve due to delayed fracture of the bulk specimen. Even if fretting action were to initiate cracks in short time, the slower crack growth rate within the bulk will delay ultimate failure of the specimen.

2.12.3.8. Contact Configurations for Fretting Fatigue Tests

Figure 2.9 shows the different types of pads and specimens that can be used in fretting fatigue experiments. Some of the pads makes a point or line contact with the specimen and usually they are used in fretting wear studies. They are called hertizian contact. They conform to the shape of the opposing surface after certain period due to wear. The present experiments on fretting fatigue were carried out in flat on flat mode of contact and many methods, as shown in Figure 2.9, are employed which modify the stress distribution at the contact area.

Ochi et. al., 2003 have investigated the effect of bridge pad geometry (Figure 2.9d) on fretting fatigue behavior. The fretting fatigue strength decreases with decrease in pad height and stick area increases with increase in pad height. The contact pressure seems to be concentrated inside the pad wherever the crack has initiated. This indicates that fretting is more pronounced when the leg height is smaller. Increase in stick implies the minimization of surface damage due to fretting and improvement in life. Bridge type of pads (Figure 2.9d) is normally used for fretting fatigue experiments which normally allows calibrating for friction coefficient.

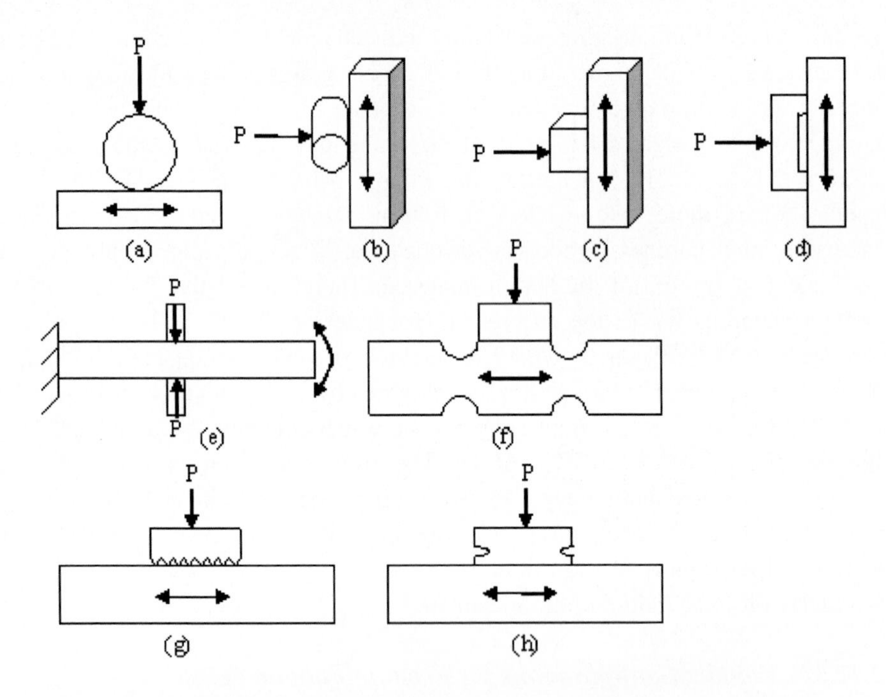

Figure 2.9. Types of specimen- pad configurations for fretting fatigue test (Hattori et. al., 2003, Kozo et. al., 2003).

2.13. Fatigue and Fretting Fatigue Studies on Titanium Alloys

Cyclic loadings are frequently experienced by the joint implants due to body motion resulting in alternating elastic, elastic-plastic or plastic deformation. Fatigue strength evaluation of implants is no doubt a difficult but critical task due to obvious differences in conditions during laboratory test and in-service conditions. The difficulty arises in correlating size, material, loading conditions of the implants. The standard material normally selected for investigation of fatigue and fracture performance of implant material is Ti-6Al-4V. These are high strength alloys used also for gas turbine parts. Its mechanical response is found to be extremely sensitive to grain size and thermo-mechanical processing history.

Widmanstatten structure is found to enhance fracture and fatigue performance. However HCF resistance has shown improvement through shotpeening. Titanium alloys are reported to be highly notch sensitive. Surface treatments such as electropolishing or porous coating for cemented prosthesis may result in reducing fatigue life by initiating cracks within the coating-substrate interface during cyclic loading. Fatigue strength also shows an improvement by the addition of interstitial elements like O, C, N and H with a slight increase in the modulus. Fatigue strength in simulated body environment has indicated varying trends. Oxygen concentration seems to play a role in altering the life of the implant although the importance of stress conditions of fatigue testing methods seems to make a difference. Corrosion resistance of Ti alloys degrades due to rupture of passive oxide films. Corrosion fatigue strength is reduced due to lower oxygen content in the simulated test solution and longer immersion time. Fretting induced corrosion fatigue is more deleterious due to enhanced electrochemical activity of the surface. The fretting motion would repeatedly expose fresh surface to the corrosive environment thereby deepening the surface damage.

Titanium alloys are also highly susceptible for fretting damage from cyclic loading action. Fretting action initiates crack on the surface and the plastic strain produced causes the disruption of surface films and leads to fatigue failure by direct metal to metal contact or third body wear from continuously produced abrasive particles. Contact temperature increases during fretting induced frictional heating generating rutile type of oxide bed within the contact. Therefore surface treatments are absolutely necessary for minimizing such surface defects on the component. Shot-peening, lubrication or surface coatings etc. are some of the most common methods employed for minimizing fretting damage. Surface oxides of titanium alloys are always very thin (in the order of 30 nanometers) and insufficient to prevent fretting failures (Antoniou et. al., 1997). Fretting behavior depends upon the type of contact material chosen. Jin et. al., 2002 has also investigated the effect of independent pad displacement on fretting fatigue behavior of Ti-6Al-4V. Fretting fatigue life which is very much influenced by slip amplitude showed minimum at 50 and 60 micron slip range and later increased with increasing relative slip range. They also remarked that debris particles generated during fretting influenced the evolution of fretting loops.

2.14. Role of Surface on the Wear Mechanism

Surface as well as subsurface characteristics plays very important role in analyzing failures due to friction and wear of interacting surfaces. Metal combination with the introduction of novel surface modification techniques has given best results in terms of minimizing friction and wear. XRD analysis of debris particles indicates the formation of thick layer of TiO over relatively thin layer of TiO_2 above the Ti matrix (Kamachi Mudali et. al., 2003). Mechanical instability of these oxide layers is quoted as responsible for increasing the friction of the contact surface. High friction on worn surfaces is generally noted by directional scratches, ridges or sharp projection for abrading and causing undue damage to the mating surface. Abrasive wear occurs when such locally rough regions plows through relatively softer surface removing the material from its wear tracks. When such particles become embedded on one of the surfaces, it further aggravates the contact layer. Therefore some amount of resistance to plastic flow of the subsurface layer is necessary to maintain the continuity of the oxide layer. This plastic flow would alter the microstructure and introduces dislocations glide along favorable crystallographic directions. Further increase in plastic strain causes dislocation pileup and cracks below the surface to induce delamination of oxides.

Chapter 3

OBJECTIVES AND SCOPE

3.1. PREFACE

The following section gives the objectives and scope of the project work on fretting wear and fretting fatigue studies on surface modified biomedical titanium alloys with respect to the current perspective of the need of such studies.

3.2. OBJECTIVES OF THE PROJECT

One of the key objectives of this project in the area of fretting wear and fretting fatigue of implant grade titanium alloys is to understand and compare the influence of surface modified layers in minimizing the surface damage during the process of fretting in a representative physiological medium. The focus of the work is also to determine the fretting damage mechanism of individual surface modified layers under fretting wear and fretting fatigue conditions.

3.3. SCOPE OF THE PROJECT

In this project an attempt is made to study the fretting wear and fretting fatigue behavior of surface modified Ti-6Al-4V and Ti-6Al-7Nb alloys. Many published articles and supplementary literatures are on hand regarding general resistance of surface modified titanium alloys towards corrosion and wear under the heading of fretting wear and fretting corrosion. But not much information is available regarding the effect of coatings on fretting behavior or fretting on fatigue behavior of surface modified titanium alloy couple. Studies on fretting behavior of coatings give good information on the quality of the coatings. Fretting fatigue studies also gives the mode of failure of specimens in presence of surface modified layers which is not investigated in depth or interpreted well so far. The same mode of failure is expected in original implants. Such a study will also help us design or modify the existing parameters of various surface modification processes for improving surface layer properties.

These alloys are treated with some of the specialized surface modification techniques such as plasma nitriding, laser nitriding, ion implantation, PVD TiN coating and thermal oxidation for minimizing fretting damage on the surface. An attempt is made to exactly simulate the conditions of load bearing implants such as hip joints and bone plates. The combination of surface modified alloys used in this work for fretting fatigue studies has not been attempted by anyone so far and it would be useful to understand the expected magnitude of friction and damage.

The basic characterization of substrate materials and surface modified layers is done with optical microscope (OM), scanning electron microscope (SEM), energy dispersive spectroscopy (EDS), roughness profilometer, nano indentation, scratch test and X-ray diffraction analysis (XRD).

Fretting wear test is conducted in air and ringer solution to understand the basic fretting response of the all the coatings. The final magnitude of damage shows the general fretting resistance of the layer.

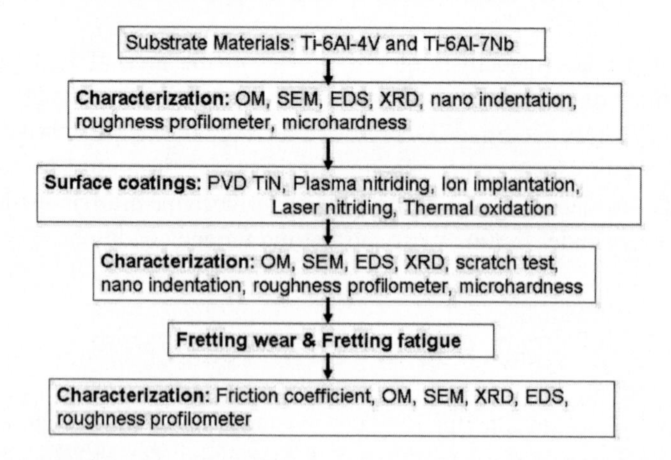

Figure 3.1. Flow chart of the experiments.

The fretting damage of individual coatings is characterized with optical microscope (OM), scanning electron microscope (SEM), friction coefficient curves, and roughness measurement. Slip between specimen and pads are measure with clip gage. Friction coefficient curves give better understanding of the progression of ongoing fretting damage. Running condition fretting response of the contact is quantified through online monitoring of friction at the contact, which is difficult to model in testing real implants. The damage sequence of individual coatings with the number of cycle can be understood better by this process of experimentation which is normally difficult in testing with original implants. It may involve much more complex sophistication and design of testing systems. Therefore, this is a kind of prototype investigation to establish the kind of coating required for implants in a particular location in the body for a particular loading condition taking into consideration the contact behavior for each coating during the process of fretting. To some extent, it can be correlated with the damage that can be expected during service of surface modified titanium alloy implants. The flow chart of the work is shown below.

Chapter 4

EXPERIMENTS

4.1. PREFACE

The following sections gives details of the materials used, the parameters used for the different surface modification processes, details of fretting wear machine and fretting fatigue fixture assembly and methods used to characterize modified layers and fretting damage.

4.2. SUBSTRATE MATERIALS

Hot rolled and annealed Ti-6Al-4V bars of dimension 200mm X 60mm X 10mm and Ti-6Al-7Nb rod of diameter 16mm and length 127mm were procured. Fatigue specimens and fretting pads were profile cut with wire-cut EDM as shown in Figure 4.1 from Ti-6Al-4V from Ti-6Al-7Nb alloys respectively. EDM cutting of alloys leaves a recast layer on the surface which is initially removed by mildly pressing on the belt grinding. The fatigue specimens were later polished with emery sheet mounted in sequence (1 – 4) on a circular shaft connected to a motor. Both the specimens and pads were polished to mirror finish with alumina slurry (3 μm) and diamond paste (0.5 μm) and later ultrasonically cleaned before the application of surface treatments. The chemical composition and mechanical properties of the alloys are shown in Table 4.1 and 4.2 respectively. Ti-6Al-7Nb alloy has more strength and ductility compared to Ti-6Al-4V alloy.

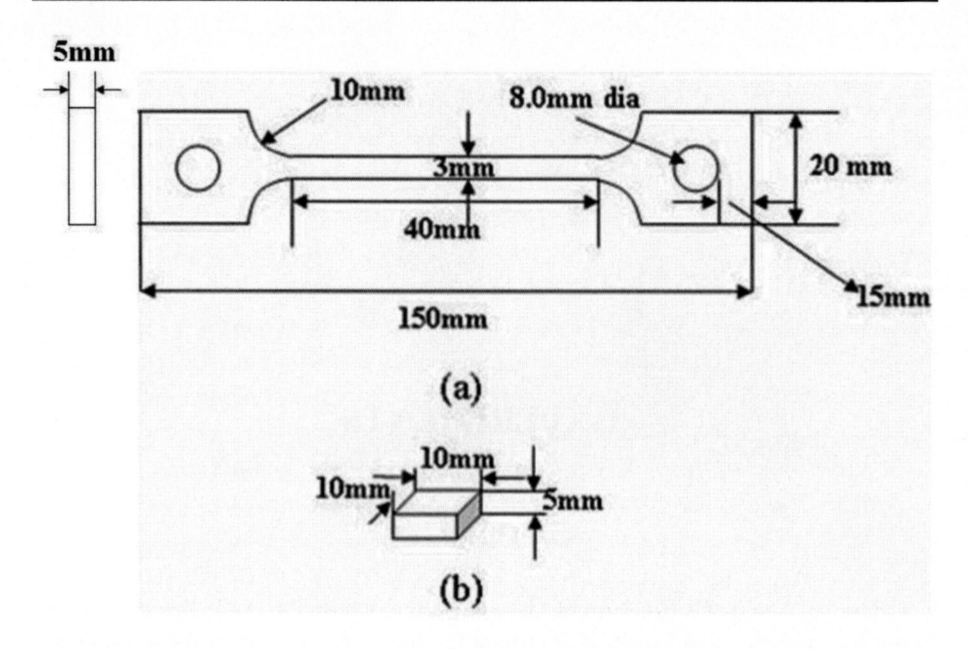

Figure 4.1. Dimensions of (a) Ti-6Al-4V fatigue specimen, (b) Ti-6Al-7Nb pads.

Table 4.1. Chemical composition of Titanium alloys (wt%).

Elements	Al	C	Fe	N	Nb	O	H	V	Ti
Ti-6Al-4V	6.21	0.016	0.05	0.0123	-	0.109	0.0025	4.16	Bal
Ti-6Al-7Nb	6.12	0.003	0.15	0.008	7.08	0.178	0.0012	-	Bal

Table 4.2. Mechanical properties of titanium alloys.

Properties	Ti-6A1-4V	Ti-6A1-7Nb
U.T.S (Mpa)	895	1027
0.2% Y.S (MPa)	825	910
% Elong	10	17
% R.A	20	48
E (GPa)	104	109
Hardness Rc	37	35

4.3. DETAILS OF SURFACE MODIFICATION PROCESSES

4.3.1. PVD TiN Coating

The specimens were initially cleaned with alumina slurry blast and later ultrasonically cleaned with acetone. The components were then placed in a vacuum chamber (316 stainless steel chamber) maintained at 10^{-4} torr. For PVD TiN coating, the pads and the specimens was placed in a vacuum chamber maintained at vacuum level of 10^{-4} torr. The samples were initially preheated to 280°C and later bombarded with Ar and H (3:1 mixture) ions to clean the surface prior to deposition. The cathodic arc process evaporates the coating material from the commercially pure (CP) titanium disc target (100 mm diameter). When this material is evaporated, a high percentage of it is ionized. An electrical charge is then applied to the substrate (~250V), which draws the ions to the surface. Nitrogen gas is steadily maintained at the pressure of 15 Pa within the chamber. The evaporated material reacts with nitrogen gas to form TiN. The process is continued until the required coating thickness (2μm) has been obtained. Then the substrate is allowed to cool and later removed from the chamber.

4.3.2. Plasma Nitriding

The specimens were placed in nitriding chamber (stainless steel) which operated within the d.c. bias voltage of 500V. The specimens were initially cleaned by surface sputtering in Ar-H_2 (3:1 ratio) plasma for 50 min prior to nitriding. The nitriding was carried out in an H_2 – N_2 (1:3 ratio) plasma for 4 hours. The pressure within the chamber is around 480 to 520 Pa. The work piece temperature is maintained in the range of 450°C to 500°C.

4.3.3. Laser Nitriding

Laser nitriding was performed with class IV 10 KW industrial CO_2 laser indigenously developed at Raja Ramanna Center for Advanced Technology (RRCAT), Indore. It is a three axis workstation with maximum laser pulse energy of 2 Joules and wavelength of 10.6 μm. The laser scan was optimized to 300mm/min and 0.4 sec interaction time. The current was maintained at 16 – 17 amps and nitrogen gas was passed from the nozzle surrounding the beam at 60 lpm for efficient transmission to the melted zone. The distance from the nozzle tip to the work piece decides the diameter of the spot. This diameter was maintained at 2mm by optimizing the working distance. The indication of nitriding is the appearance of blue plasma during melting and golden orange tinge at the nitrided area. Laser nitriding induces many surface undulations with brownish ripples since the modified layer is deeper than other process due to melting and resolidification. Therefore, they were grinded with alumina wheel for removing the uneven part.

4.3.4. Ion Implantation

Ion implantation is conducted with TAMSAMES 150 keV particle accelerator machine. Both the pads and the specimens were cleaned with acetone and fixed inside a cylindrical stainless chamber with a circular channel window at the side for sending ion beam towards the target. The specimen area to be implanted is exposed towards the incoming beam. The dose of implantation is fixed to 2 X 10^{16} ions/cm^2 with the energy of 100 keV. The beam current was maintained within 2μA and the sample temperature was below 40 °C. The vacuum was maintained below 10^{-7} torr throughout the test. The time for implantation of single area was more than 5 hours. This makes the process more expensive and time consuming.

4.3.5. Thermal Oxidation

The specimens were thermally oxidized at 650 °C in a tube furnace in ambient atmosphere for 72 hours. The oxidized layer consists of black rutile (about 100 μm) above the hard alpha case (about 20 μm). Rutile was removed meticulously with emery paper without removing the case.

4.4. Fretting Wear Machine Details and Testing Methodology

The basic components of test rig used for fretting wear studies are as shown in Figure 4.2a. The details of the machine are given elsewhere (Ramesh and Gnanamoorthy, 2006). The original mating components undergoing fretting wear damage are prepared into specimens and butted against each other (1 & 2). A reciprocating system based on motor is designed to induce fretting oscillation of desired amplitude (3). Vertical motion along one end of the machine is translated to horizontal motion at the sample holder. Reciprocating rod is butted to sample holder via a force transducer. Frequency is basically decided by the motor speed (r.p.m). Fretting amplitude is manipulated by adjusting the cam connected to the motor. A feedback circuit system senses the relative displacement. Dead weights (4) are normally used to apply the contact pressures. Ball-on-flat (point contact) system was used in the present experiments. The tangential force and sliding amplitude is sensed with calibrated S-beam force transducer and displacement sensors (5 & 6). Displacement sensors usually come with non contact laser pointer system or eddy current sensing system. A laser beam pointer sensor with 1μm accuracy was used to record relative displacement in this machine. Tangential force is continuously monitored and recorded to explain the contact behavior during fretting.

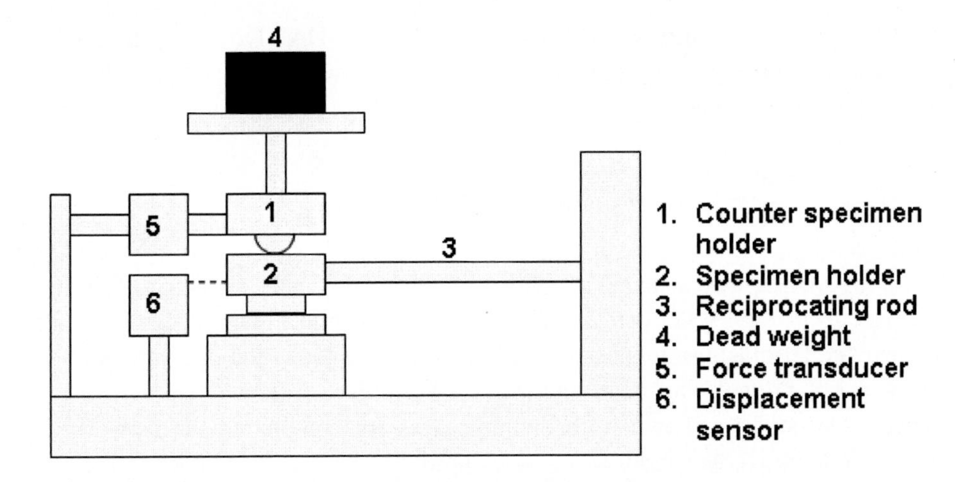

Figure 4.2. Test rig for fretting wear studies (Ramesh et. al., 2006).

Alumina ball of 10mm diameter was chosen as the counter material on flat specimens (10mm X 10mm). Both the specimen and the ball were cleaned with acetone before making the contact every time. Dead weights were used to apply normal pressure. The normal force applied in all the tests was 1N. Fretting frequency was adjusted to 5Hz at sliding distance of 200 μm. This requires manipulation of motor driven reciprocation system. Tangential force and displacement was continuously recorded through data acquisition system.

The specimen holder is designed to contain ringer solution to conduct fretting experiments within the liquid environment. Ringer solution with the following composition was used for both fretting wear and fretting fatigue experiments: NaCl (6.5 g/l), KCl (0.14 g/l), $CaCl_2$ (0.12 g/l), $NaHCO_3$ (0.2 g/l) and Dextrose (0.4 g/l). All the experiments were conducted both in ambient atmosphere and in ringer solution. The relative humidity (RH) was 85% during the test in air. Humidity level also affects the friction at the contact. Hardened steel balls (10mm diameter) were also used to generate fretting loops for uncoated alloys at different loads. Two tests were conducted for all the conditions to check the consistency of the work.

4.5. Fretting Fatigue Test

4.5.1. Machine Details and Test Rig Design

Fretting fatigue tests were conducted in 100 kN DARTEC servo-hydraulic UTM with 9600 Hydrowin modular. The specimens were held at two ends with cotter-pin self aligning fixtures.

4.5.2. Designing of Proving Ring and Pad Holders

Proving ring was designed to apply the lateral force on the specimen. The ring is made from AISI 4340 (En 24) alloy steel as shown in Figure 4.3. Four strain gages with terminals were pasted at the opposite ends along the circumference to form a full bridge Wheatstone network. Two active gages were provided for sensing the strain and two were dummy gages for completing the network as shown in Figure 4.3a. Circuit wires were appropriately soldered to the terminals.

Two similar pad holders were profile cut with wire-cut EDM form 316 stainless steel as shown in Figure 4.3b. Strain gages were pasted underside of both the pads. The pads were fixed rigidly with Allen screws after ensuring proper flatness of contact on the specimen. The alignment of pads on fatigue specimen is obtained automatically from butting hardened steel balls on a pit drilled behind the pad holders. Stainless steel fixtures were also deigned to hold the pads rigid during the fretting fatigue test.

Five channel strain amplifier was used to record the pad strains. One full bridge channel is used for proving ring and two half bridge channels for two pad holders. Pad holder strain gages were clubbed in series with a resistor of similar value. This makes the half bridge complete to couple with in-built resistors within the stain amplifier. Strain amplifier output is connected to the computer through RS232 cable and the values were obtained and recorded in hyperterminal.

4.5.3. Calibration of Proving Ring and Fretting Pads

Proving ring was calibrated according the ASTM E-74 – 02 standard, which gives the standard practice for calibrating force measuring instruments and verifying the force indication of testing machines.

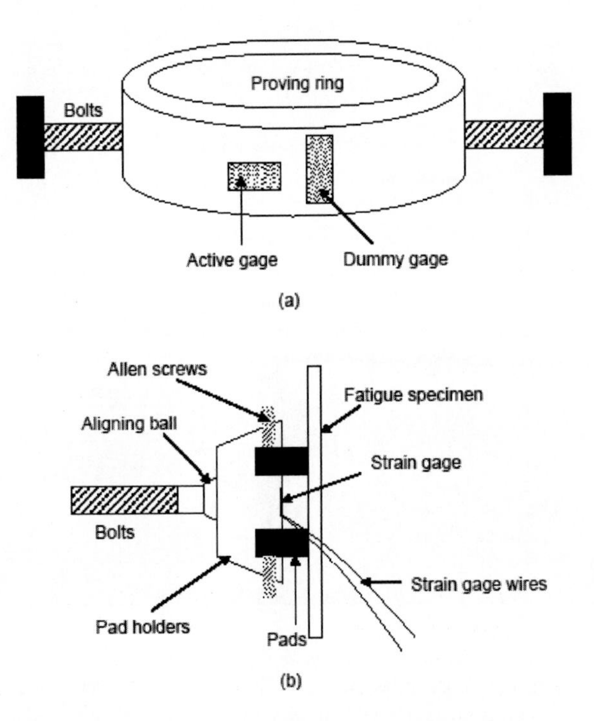

Figure 4.3. Schematic sketch of design of proving ring and pad holder.

The calibration involves applying a known load to the ring and recording the strain values. The strain generated during the circumferential deformation of the ring is proportional to the applied load. The ring was held rigidly in DARTEC universal testing machine along the bolts and a series of tensile load is applied. Then the strain values corresponding to the load is recorded. This procedure was repeated several times to get a steady value. Pad holders were similarly calibrated for each of the surface coatings. Each specimen were cut into two halves and assembled with a small gap between them. The pads and the proof ring were then assembled. A small sinusoidal displacement is applied to the specimen under stroke mode and the transmitted load is recorded as a function of pad strain. The slope of this graph gives the pad calibration value which enables the conversion of measured fretting pad strain to frictional force.

4.5.4. Fretting Fatigue Test

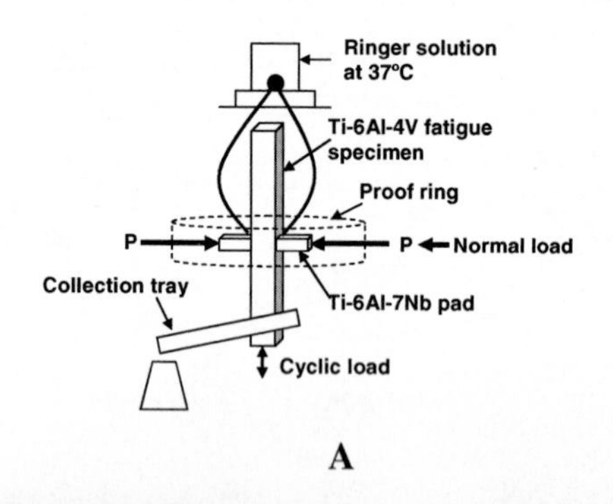

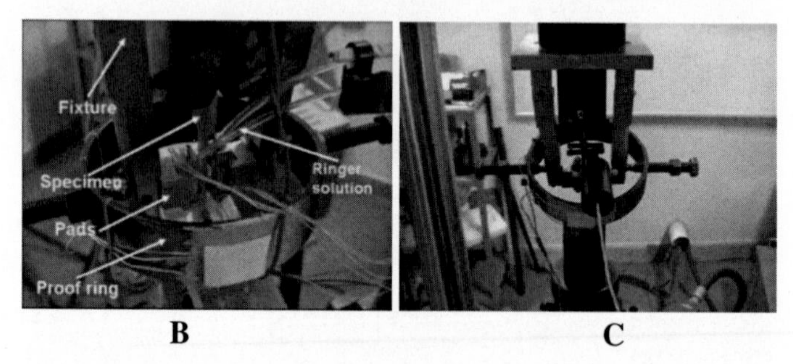

Figure 4.4. (a) Experimental setup for fretting fatigue test, (b) Camera view of fretting fatigue setup, (c) Slip amplitude measurement setup with CTOD clip gage.

Contact pressure is applied by tightening the bolts along the circumference of the proving ring (Figure 4.4 a & b) and tangential force is deduced and recorded from strain gages bonded below the pad holders. Modular junction of the hip shafts normally experiences two to three times the weight of the upper portion of body above the hips (200 to 600 kilograms) (Kamachi Mudali et al (2003)). It would roughly correspond to 40 MPa although it varies with design of the prostheses and load excursions, like climbing staircase up or down sometimes causing sudden impact along the taper section of the joints. Therefore, contact pressure of 40 MPa (600N) was applied before testing all the specimens. Cyclic loads of 3 to 7.5 kN at 5 Hz frequency were applied during the test. Stress ratio R was maintained at 0 because the modular junction undergoes tensile force at the contact point while compressive load is applied on the ball.

Slip range between pads and the specimens were measured with CTOD clip gage as shown in Figure 4.4c.

Ringer solution contained in a glass jar maintained at 37°C was continuously streamed on pads with a rubber hose and collected in a container below. This flow is adjusted to obtain complete immersion of contact area similar to fretting action of hip joints surrounded by continuous flow of physiological medium.

4.6. Characterization Techniques

4.6.1. Microscopic Examination

Microstructural examination of substrate materials, surface modified layers and fretting damage areas is carried out with Zeiss MC43 optical microscope and JSM - 840A JEOL Scanning electron microscope (Japan). EDS analysis of the damaged area is done with ISIS link microanalysis, Oxford Instrument – UK. The etchant used is Keller's reagent (6 ml HNO_3, 2 ml HF and 92 ml distilled water).

4.6.2. Microhardness Measurement

Vickers microhardness profile measurement of plasma nitrided and thermal oxidized titanium alloys is done with Leco M-400 microhardness tester. The load level selected ranged from 5gms to 500gms. The hardness was measured for every 2 micron interval from the surface and also the load carrying capability of modified layers is done by measuring hardness with increasing load from 25gms to 2000gms. The test was repeated three times to check consistency.

4.6.3. X-Ray Diffraction (XRD) Analysis

XRD analysis is performed on uncoated as well as coated specimens with Shimadzu XD-D1 (Cu K_α) X-Ray diffractometer. A small piece of substrate and coated samples were always preserved for microstructural as well as XRD tests.

The diffraction peaks were indexed with JCPDF software. The same is compared with the other journal publications for confirmation.

4.6.4. Scratch Test

Scratch testing of PVD TiN coating, plasma nitriding and ion implanted alloys is done with custom-built scratch test machine. A Rockwell stylus was fixed normal to the samples. The stylus traverse at the speed of 2mm/sec on the sample fixed to a rigid holder. The sample holder was tilted to one degree which applies progressively increasing normal load with the movement of the stylus. The sudden drop in the normal and traction force indicates the rupture strength and adhesion strength of the coating. Each test was repeated two to three times to check for consistent values.

4.6.5. Nano Indentation

Nano indentation test is performed with Nano X40NM to measure the hardness and reduced elastic modulus of thin coatings. The samples were cleaned with acetone and glued to a small steel disc and placed on a magnetic holder. The indentation area is initially selected after observation with optical microscope. The Berkovich indenter later scans a small part of the surface with higher depth of focus similar to AFM. The indentation area can be selected from the scanned picture. Normal load applied ranged from 100μN for ion-implanted samples to 500 mN for unmodified sample. The force and displacement resolution of the instrument was below 0.04 μN and 0.04 nm respectively. The load-depth curve is plotted from the values recorded during indentation. The hardness and the elastic modulus of the surface layer is calculated from the maximum load and unloading part of the curve respectively. The control software automatically shows the values of hardness and elastic modulus. Test was repeated three times to get consistent values.

4.6.6. Roughness Measurement

The roughness profiles of coatings and fretting damage were measured with M2 stylus profilometer. The travel distance was varied from 1.6 mm to 5.4 mm.

Chapter 5

RESULTS AND DISCUSSION

5.1. PREFACE

This section is described in four different parts beginning with characterization results of substrate materials and surface modified layers followed by results and discussion on fretting wear and fretting fatigue tests of surface modified alloys. The final section describes some of the important features of fretting fatigue failures commonly observed during the tests. The references to the results described below are very limited since the work is entirely a new approach in the area of biotribology.

5.2. CHARACTERIZATION OF SUBSTRATE MATERIALS

5.2.1. Microstructural Details of Substrate Titanium Alloys

Figure 5.1 shows the optical micrograph of Ti-6Al-4V and Ti-6Al-7Nb. The volume fractions of primary α in Ti-6Al-4V and Ti-6Al-7Nb alloys are 53% and 91% respectively. The volume percent of α+β colony in Ti-6Al-4V and Ti-6Al-7Nb alloys is 47% and 9% respectively. The lamellar α+β colony of Ti-6Al-4V is more resolved compared to Ti-6Al-7Nb with smaller primary α size. Since the alloy is hot rolled and annealed, it shows bimodal structure with equiaxed primary α within α+β matrix. Microscopic observation along the longitudinal direction (rolled surface) reveals elongated lamellar structure. Bimodal structures are generally known for its high ductility and fatigue strength with well balanced properties due to combined influence of lamellar and equiaxed microstructures. In these alloys, Al stabilizes h.c.p alpha phase and Nb and V stabilizes b.c.c beta phases in these alloys. Figure 5.2 shows the XRD analysis pattern of the base materials indicating the presence of α phase. The surface of Ti-6Al-4V and Ti-6Al-7Nb will be dominated with semicrystalline TiO_2 layer (~30 nm) along with oxides of Al, V and Nb respectively (Sitting et. al., 1999). Fatigue lives are generally governed by the volume fraction of the phases present. The presence of finer phases on the surface of unmodified materials is favorable for increasing fretting fatigue crack initiation life.

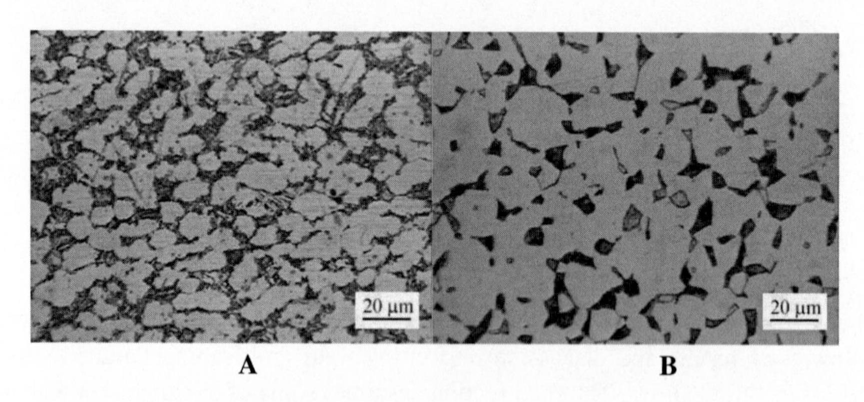

Figure 5.1. Optical micrographs of (a) Ti-6Al-4V, (b) Ti-6Al-7Nb.

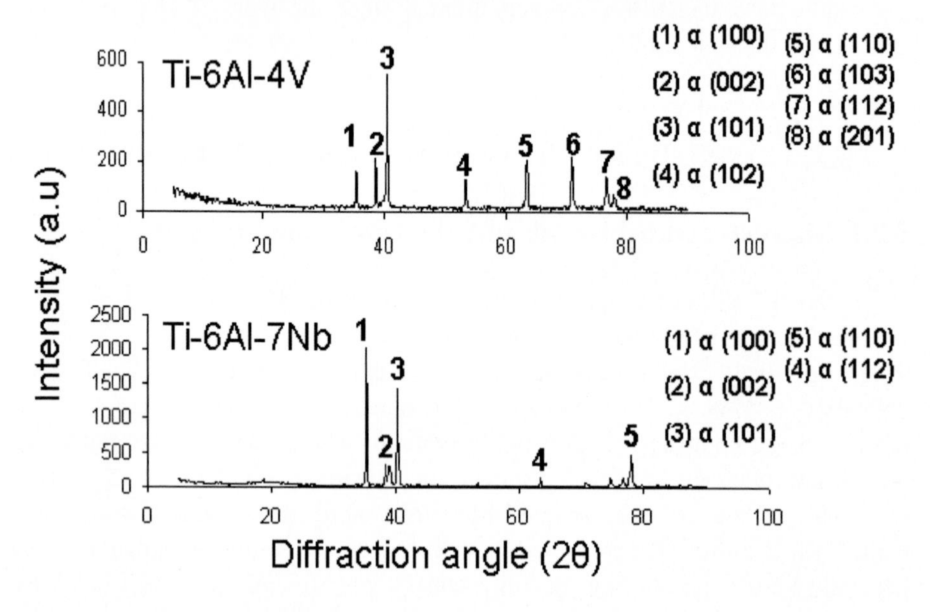

Figure 5.2. X-Ray Diffraction pattern of substrate titanium alloys showing α phase.

5.2.3. Nano Indentation Test

Figure 5.3 shows the load-depth nano indentation plot of Ti-6Al-4V substrate. Nano indentation is suitable for studying the properties of thin films or modified layers. It can be used for all possible materials in engineering applications. Normal load is applied gradually through a diamond indenter (Berkovich indenter) which will penetrate few tens of nanometers from the surface. The load and depth are continuously recorded as shown in Figure 5.3. For Ti-6Al-4V substrate, maximum normal load of 5 mN was applied and the penetration depth was below 60nm. The surface layer properties are calculated with constitutive equations and presented in terms of hardness and reduced elastic modulus. The elastic modulus is reduced due to small amount of deformation experienced by the indenter during penetration. The hardness and reduced elastic modulus of the substrate are 4 GPa and 129 GPa respectively. Nano indentation test results of all thin layer coatings are presented in the subsequent sections.

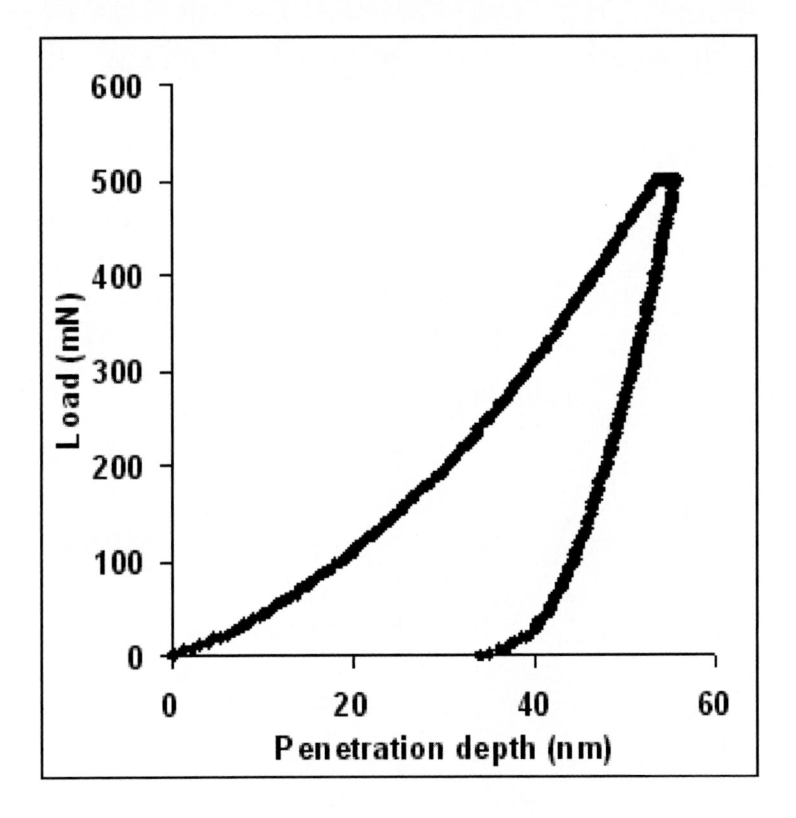

Figure 5.3. Nano indentation profile of Ti-6Al-4V substrate.

5.3. Characterization of Surface Modified Layers

5.3.1. PVD TiN Coating

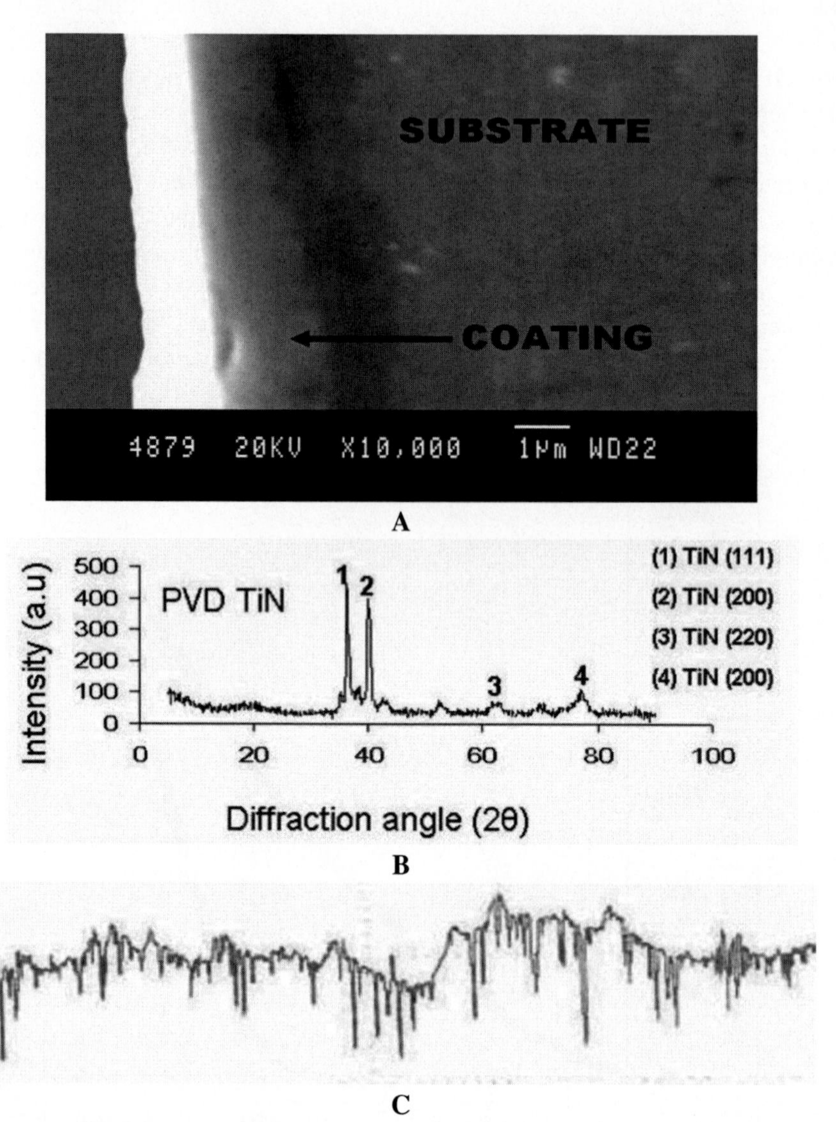

Figure 5.4. (a) PVD TiN layer on Ti-6Al-7Nb, (b) XRD of TiN layer, (c) Surface roughness profile of PVD TiN coated alloy ($R_a = 0.3$ μm)

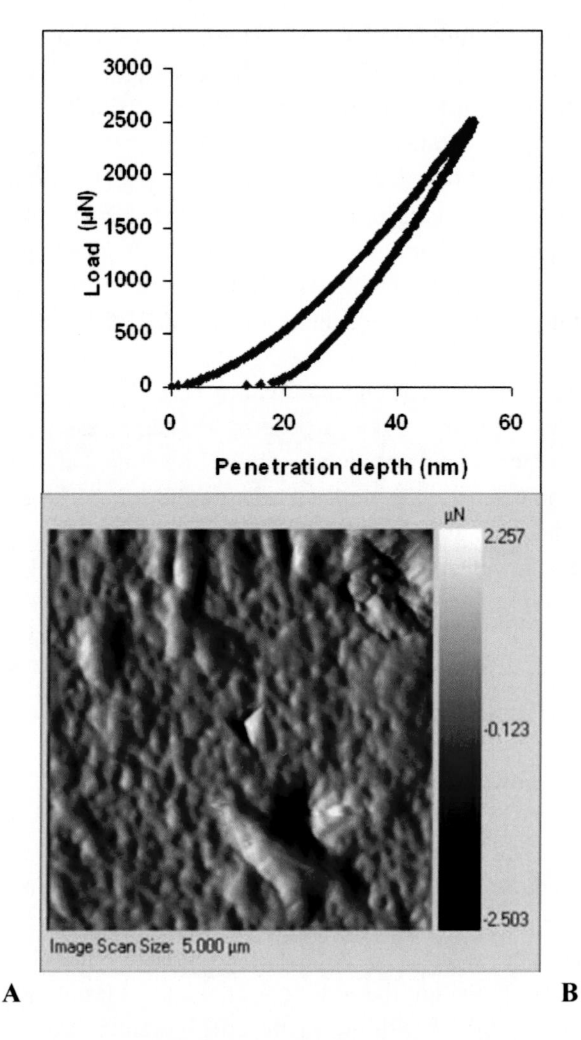

A B

Figure 5.5. (a) Nano indentation profile of PVD TiN coating, (b) Surface profile.

Figure 5.4a shows coating of TiN on Ti-6Al-7Nb and Figure 5.4b shows the XRD pattern of the TiN layer indicating the pure titanium nitride phase. The growth of TiN layer is normally columnar on the substrate during the PVD process. This feature however was not resolvable in the presented micrograph. The coating introduces bi-axial residual stresses, which scales with the substrate yield stress, and strongly resists the propagation of shear fracture along the columnar boundaries (Bhowmick et. al., 2004). Therefore it appears that the integrity of the TiN columns is mainly due to the compressive residual stresses developed during the deposition. The main deformation mechanism of TiN during indentation is shear fracture ruling out any possibility of dislocation slip effecting gross plastic deformation. The hardness of the layer is 2800 HVN with adhesion strength of 90N. The surface feature and roughness profile (R_a = 0.3 μm) as shown in Figure 5.4c indicates uneven surface of the coating. Lots of debris can be expected during fretting within the contact from the interaction of surface projections. It is also sometimes assumed favorable during fretting because the surface depressions can provide sink for the generated particles thereby minimizing three body wear. Finally the ability to resist exfoliation of the film from the substrate depends on the adhesion strength, purity and uniformity of the film.

Figure 5.5a shows the nano indentation profile of PVD TiN layer. Maximum load of 2.5mN was applied to get a penetration depth less than 60nm. Elastic recovery is much more compared to unmodified substrate due to high hardness of the layer. It is indicated by the reduced area of the loop. The hardness and the reduced elastic modulus are reported as 40 GPa and 332 GPa respectively. Figure 5.5b shows the nano indenter scanning probe micrograph of the layer with the indenter impression.

5.3.2. Plasma Nitriding

Figure 5.6a shows the cross section of plasma nitrided Ti-6Al-4V alloy. Plasma nitriding deals with thermal diffusion of nitrogen under negative pulsed DC biased voltage and temperature. The gradient or diffused layer is more clearly visible in the picture. The case thickness is dependent upon the processing temperature and time. XRD results as shown in Figure 5.6b indicate the formation of two prominent phases TiN and Ti_2N. These phases are not clearly demarcated since the layer must be within the order of nanometer. Plasma nitriding at higher temperatures (700 – 900°C) with higher nitrogen potential produces thicker case with larger volume fractions of TiN and Ti_2N. Thicker and harder modified case is not at all preferable because it always favors reduction in fatigue strength (Ani Zhecheva, 2004). The weight fractions of these phases will normally vary within the nitrided layer. TiN is always found in higher volume towards the surface due to availability of higher amount of nitrogen (Bekir et al, 1996). The layer beneath TiN would be a mixture of Ti_2N and TiN. Solid solution of nitrogen in α is found below the mixed layer. Therefore there is a gradient of hardness below the surface layer. In such case, the gradient layer may serve double purpose of preventing catastrophic failures due to thicker and harder modified case while delaying the fretting induced surface defects. The softer volume of the layer will have more capacity to accommodate the surface and subsurface plastic deformation due to fretting. However, the layer is comparatively stronger than the base substrate. This will ensure higher fretting fatigue strength compared to unmodified alloy.

The plasma nitrided samples are smoother (R_a = 0.035μm) than as polished unmodified alloy (R_a = 0.113μm). Since the layer is chemically inert due to the formation of ceramic nitride compounds, the friction will be minimum compared to unmodified alloys.

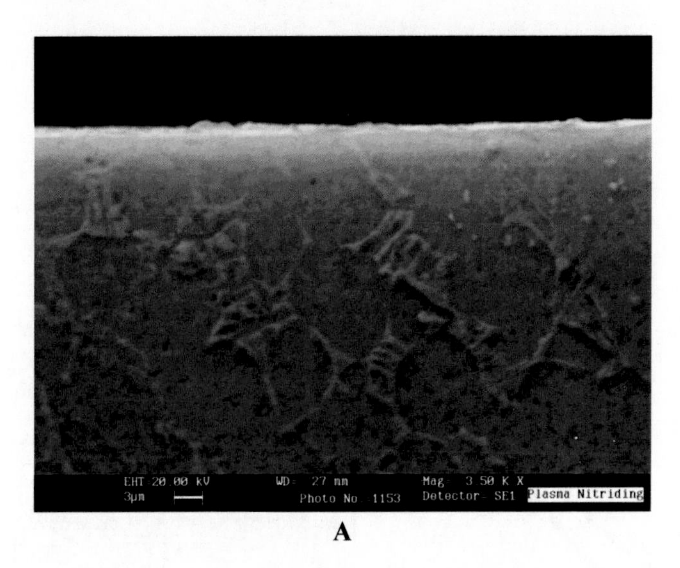

A

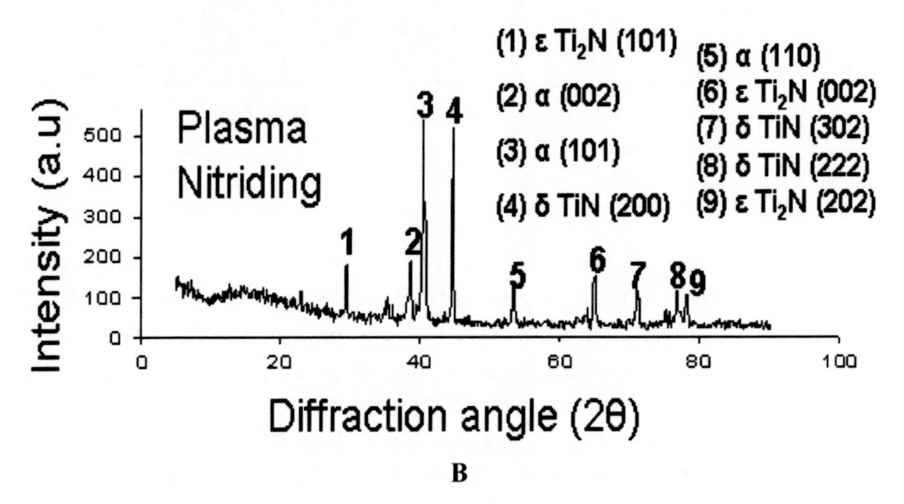

B

Figure 5.6. (a) Plasma nitrided layer, (b) XRD pattern of the nitrided layer.

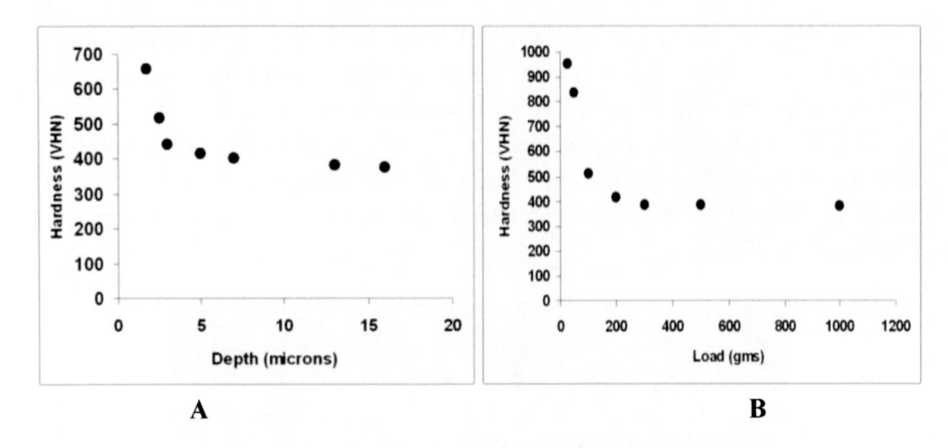

Figure 5.7. (a) Hardness depth profile and (b) Hardness load profile of plasma nitrided Ti-6Al-4V.

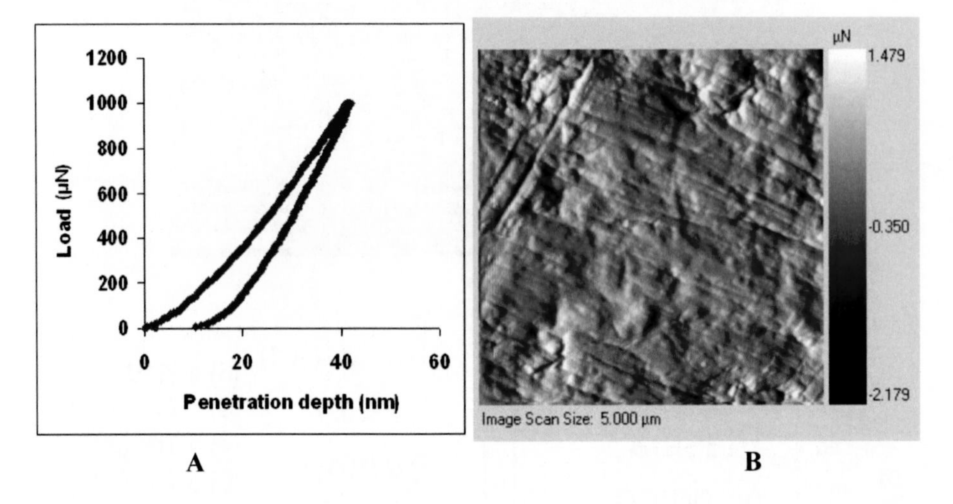

Figure 5.8. (a) Nano indentation profile of plasma nitrided Ti-6Al-4V, (b) Surface profile.

Figure 5.7 (a) and (b) shows the hardness profile of the plasma nitrided Ti-6Al-4V alloy. It shows a decreasing trend from the surface indicating the reduced concentration of nitrogen below the surface. It can be observed that the hardness is below 700 VHN within at a depth of 2 µm. The hardness produced by plasma nitriding process usually crosses 1000 VHN on the surface. This indicates that hardest layer with TiN and Ti_2N is less than 2 µm. This effect can be observed in Figure 5.7b. The hardness drops after certain load when the indenter penetrates the layer giving the composite hardness rather than the film hardness. The gradient layer may be more useful during fretting fatigue because the surface strains developed during fretting can be easily relieved by softer layer than hard coatings like PVD TiN which may experience failure due its inability for accommodating fretting induced tangential strains. This fact is discussed later. Figure 5.8a shows the nano indentation profile of plasma nitrided Ti-6Al-4V alloy with maximum load 1 mN giving penetration depth of about 40nm. The layer is softer compared to PVD TiN layer with hardness and reduced elastic modulus of 23 GPa and 201 GPa respectively. Figure 5.8b shows the nano indenter scan probe micrograph of the surface showing fine scratches.

5.3.3. Nitrogen Ion Implantation

The ion implanted specimens do not generally produce contrasting changes in surface color like thermally oxidized or PVD TiN coated specimens. However, a slightly burnt brownish appearance was visible indicating the implantation. From the TRIM (Transport of Ions in Matter) software, the projected range of nitrogen in this titanium alloy was found to be 170 nm. This is the case because, the ions are forced to diffuse within the layer with the acquired kinetic energy without the aid of any other force. The ions trajectory therefore depends upon the mass and the kinetic energy.

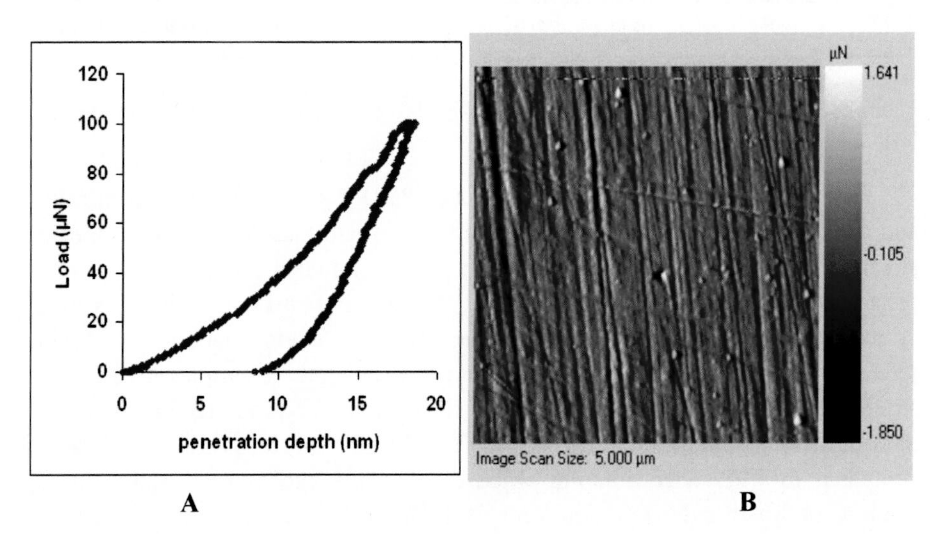

Figure 5.9. (a) Nano indentation profile of nitrogen ion implanted Ti-6Al-4V (b) Surface profile.

Figure 5.9a shows the nano indenter profile of nitrogen ion implanted titanium alloy. Maximum load of 0.1 mN or 100 μN was applied since the layer is very thin. The penetration depth is below 20 nm, which is well within the surface modified layer of 170 nm. The hardness and the reduced elastic modulus of the layer are reported as 6 GPa and 115 GPa respectively. Figure 5.9b shows the nano indenter scan probe micrograph of the surface of ion implanted alloy. The indenter mark is also seen at the center.

Figure 5.10 shows the comparison of the properties of all the thin layers. This shows that PVD TiN layer has the highest hardness and the elastic modulus among all the thin layers. Ion implanted samples have shown least improvement in the properties.

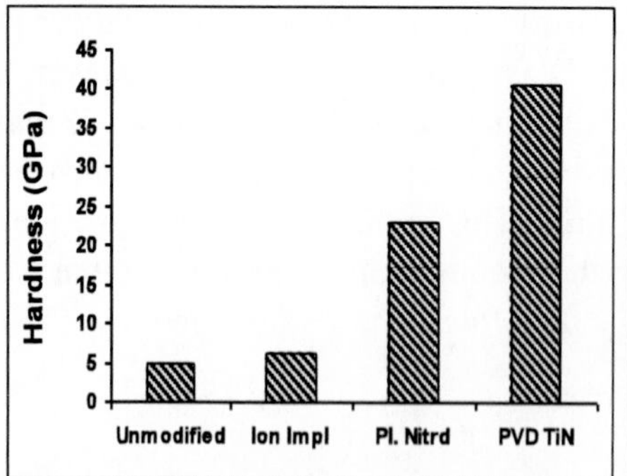

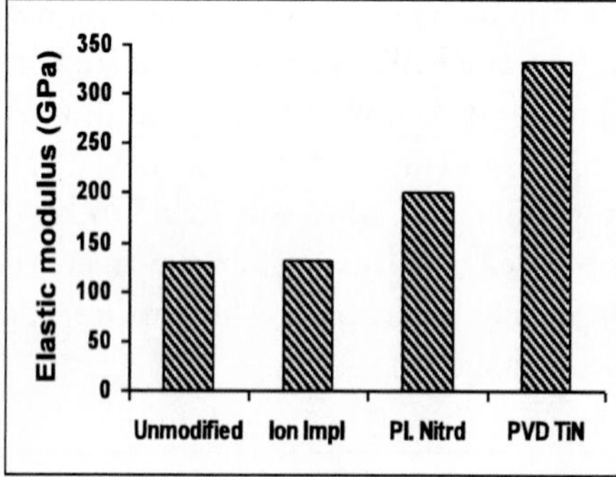

Figure 5.10. Comparison of the properties obtained from the nano indentation test of thin layers.

Figure 5.11 shows scratch test results of PVD TiN coating, plasma nitriding and ion implanted layers. Scratch testing is normally used to obtain adhesion strength of the surface coated layers. It can also indicate the general scratch resistant property of modified layers. PVD TiN has better strength in terms of critical rupture and traction force compared to other layers. Traction force value refers to adhesion strength of the layer to the substrate. Therefore it cannot be strictly compared with other processes since all other processes used are the modification techniques except PVD TiN. Plasma nitrided alloys are softer compared to PVD TiN. Ion implanted alloys have shown least value due to shallow modified layer. Traction force of modified layers indicates the resistance to plowing action of the diamond indenter.

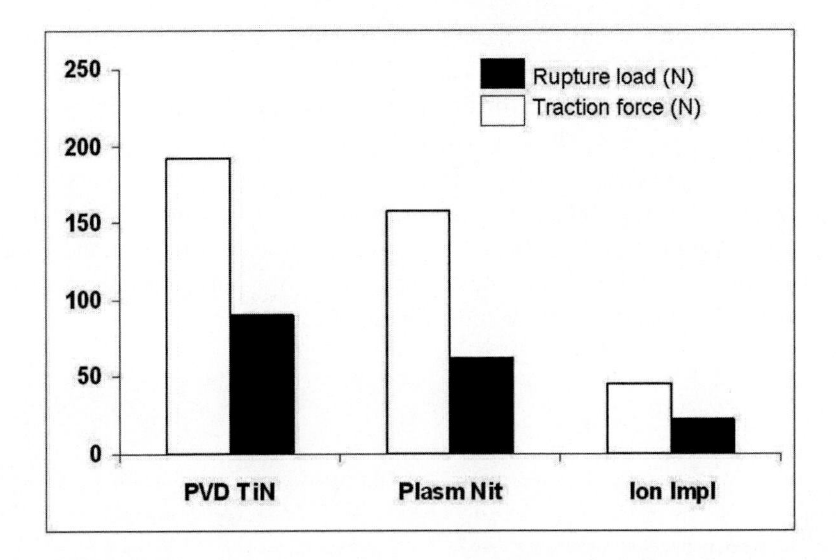

Figure 5.11. Scratch test results of thin layers.

5.3.4. Laser Nitriding

Figure 5.12a shows the gage portion of laser nitrided Ti-6Al-4V fatigue specimen. It appears more like a weld overlay along the load axis of the specimen. The edges have melted due to high power density of the laser. The depth of penetration is as shownin Figure 5.12b, is extremely high compared to all other process used in these experiments. Undercuts and cracks were also observed along either sides of extended overlay portion on the flat grip area of the fatigue specimen. Macroscopic observation reveals golden brownish melt deposit indicating the formation of nitrides. The hardness of TiN deposit ranged from 1450 to 1600 VHN. The scan speed and the power density were optimized with few test coupons to obtain minimum penetration and complete nitridation. The optimized power density should provide minimum penetration and complete dissociation of nitrogen at the melting zone. The appearance of blue plasma during melting is the indication of dissociation of nitrogen. One of the recommended methods offered was to have high power density with minimum interaction time to have complete dissociation of nitrogen as well as minimum penetration. This will also offer the advantage of developing finer microstructural features on the surface. The nitriding can also be made efficient through controlled high purity nitrogen atmosphere around the material. Convection currents in combination with high level of diffusion of nitrogen in the melt ensure complete nitridation. Figure 5.13 shows the boundary between base material and the HAZ. Grain size difference is clearly visible between the substrate and HAZ. HAZ grains are finer and barely distinguishable even after deep etching. Geeta et. al., 2004 indicates that needle like structure just below the melt deposit in Figure 5.14 consists of mixture of $TiN_{0.3}$, TiN and α' - Ti. Dendritic TiN is within the melted zone and hexagonal $TiN_{0.3}$ is also formed close to the melt/HAZ boundary as shown in Figure 5.14. The boundary between brittle melt deposit and HAZ is sharply outlined which can become a potential site for initiating failure because the properties of the layer and the substrate significantly vary.

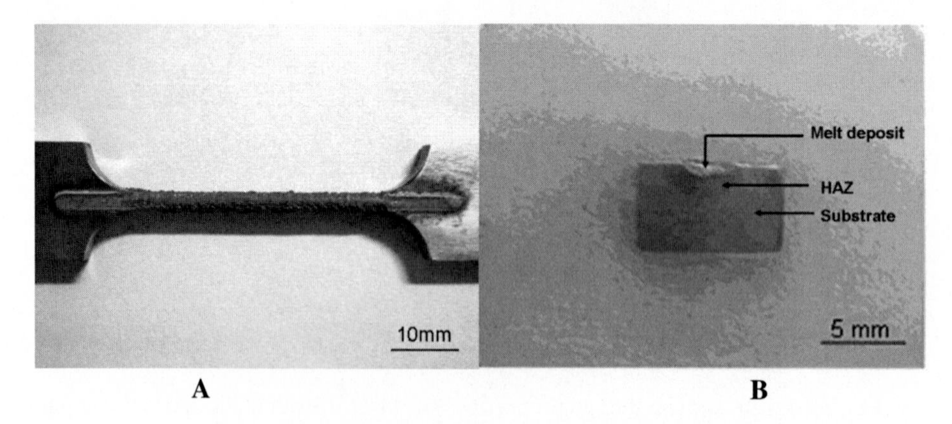

Figure 5.12. Laser nitrided Ti-6Al-4V (a) Fatigue specimen, (b) Sample cross section.

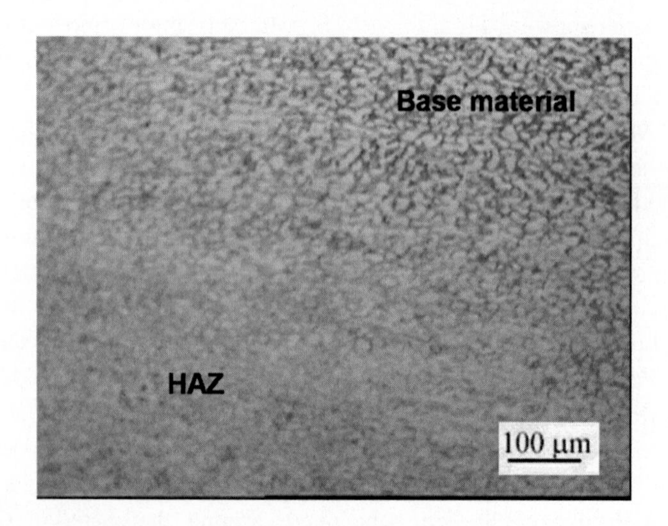

Figure 5.13. HAZ-base material boundary.

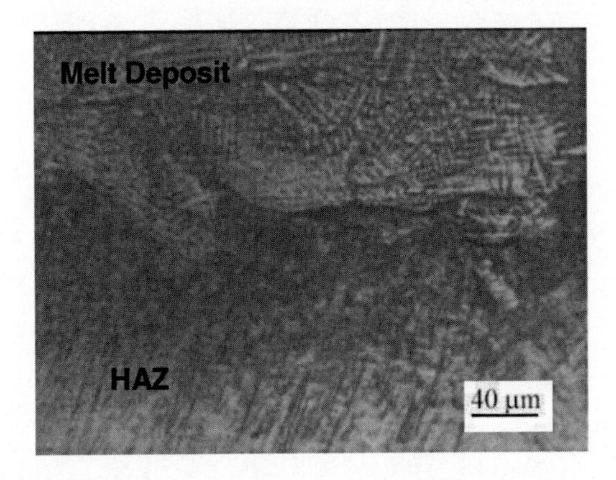

Figure 5.14. Melt deposit - HAZ boundary.

Figure 5.15 shows the microstructural features of laser nitrided Ti-6Al-4V alloy. The melt deposit is almost 0.3 to 0.5 mm thick consisting of randomly oriented TiN dendrites (60 to 70 Vol %) as shown in Figure 5.13a. Dendritic features are expected due to melting of substrate, alloying and resolidification similar to a mini casting process. Figure 5.16 shows XRD analysis of laser nitrided Ti-6Al-4V confirming the presence of TiN and Ti_2N phase.

The volume of dendrites can also be varied by dilution of nitrogen with inert gases. Figure 5.15b shows the heat affected zone (HAZ) adjacent to melt deposit. This is a typical martensite α' structure due to faster cooling of the region. It is interesting to observe that the growth of needles is perpendicular to each other in certain favorable crystallographic orientation. The average hardness of HAZ is around 432 VHN which is slightly more than the base material.

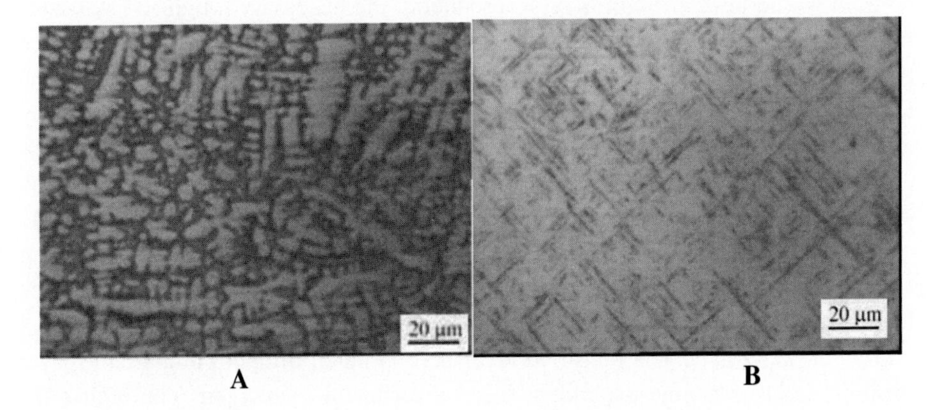

Figure 5.15. Microstructural features of laser nitrided Ti-6Al-4V (a) Melt deposit, (b) Heat Affected Zone (HAZ).

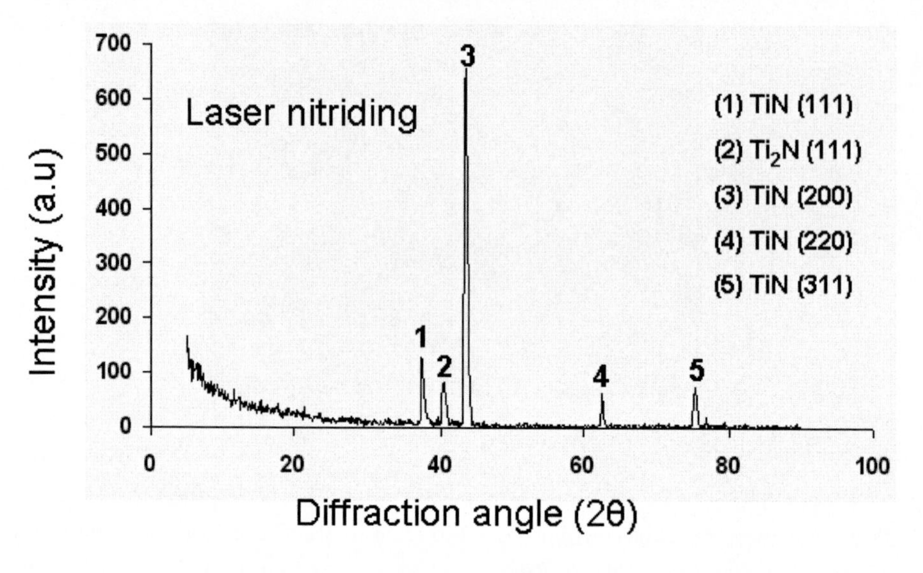

Figure 5.16. Diffraction pattern of laser nitrided Ti-6Al-4V.

5.3.5. Thermal Oxidation (To)

Thermal oxidation of titanium alloys has produced very rough surface with black oxide layer on the top and hard α case (solid solution of oxygen in titanium in high concentration) beneath the oxide layer as shown in Figure 5.17. The interface between black oxide and α case is highly irregular as observed in Figureure. In such case, removal of oxide layer without disturbing the case beneath would be a difficult task. Reproducing the necessary roughness will also be a problem. Usually there will be a gradient of hardness across the layer since the oxygen concentration is not uniform through out the layer cross section. XRD analysis shown in Figure 5.18 confirms the rutile phase formation (indicated by R) due to high temperature oxidation. Figures 5.19 (a) and (b) shows the hardness profile of the thermally oxidized layer which shows a decreasing trend from the surface.

An oxidized titanium alloy has higher strength and lower ductility on the surface due to stabilization of α phase. It is reported that α case within the surface layer of titanium improves tribocorrosion and galling resistance of titanium alloys (Hasan et. al., 2004). The modified layer here is much thinner compared to laser nitriding, as it is a time dependent thermal diffusion of oxygen. The higher the temperature and time, the higher will be the case depth obtained. Thicker and harder case will degrade the fatigue properties. The case depth can be controlled with time of oxidation. Controlled oxidizing atmosphere can also be used for thermal oxidation process.

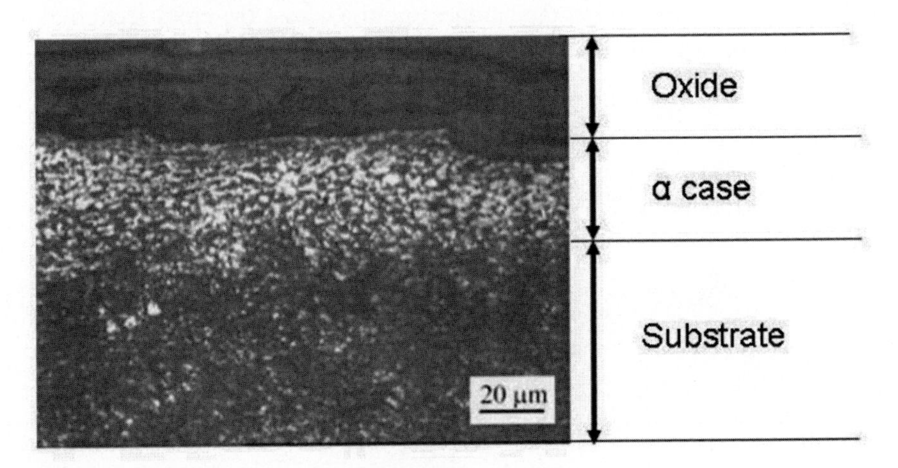

Figure 5.17. Thermally oxidized Ti-6Al-4V showing the oxide layer at the top and α case beneath the oxide layer.

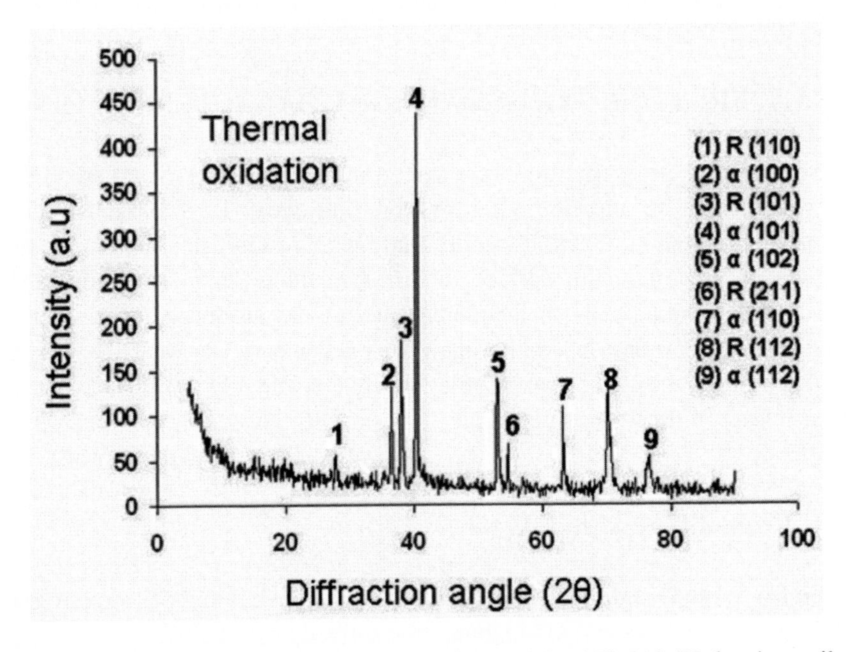

Figure 5.18. X-Ray Diffraction analysis of thermally oxidized Ti-6Al-4V showing rutile and alpha phase.

Figure 5.19a shows the load carrying capacity of the layer. At lower loads, the layer offer good indentation resistance and at higher loads, the indenter penetrates the layer. Figure 5.19b shows the hardness profile across the cross section of the sample. Hardness gradient is proportional to concentration gradient across the layer.

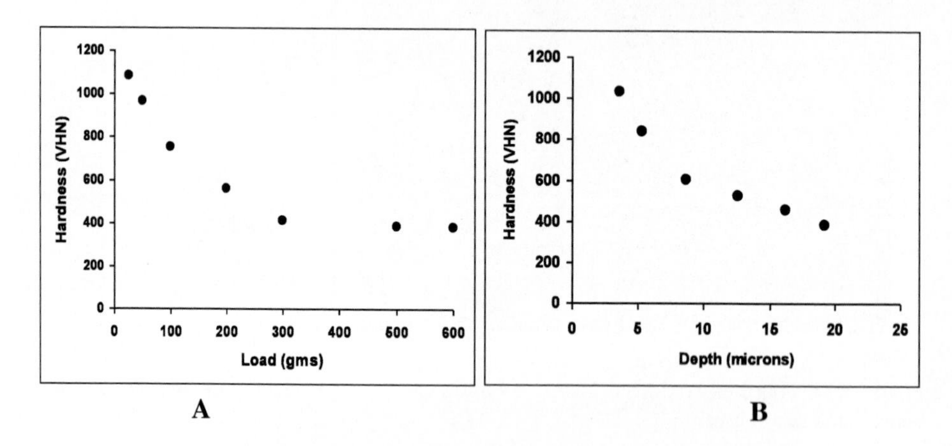

Figure 5.19. Microhardness profile of thermally oxidized Ti-6Al-4V.

5.4. Fretting Wear Damage Characterization

This section gives fretting wear test results on surface modified titanium alloys compared to unmodified alloy. Fretting wear is a possible damage mechanism in modular junctions of hip shaft models comprising of ceramic heads (alumina or zirconia) with surface modified metallic shafts (titanium alloys). Fretting with alumina balls, as a counter material, has produced a wear scar with different configurations depending upon the parameters. The wear rate has been calculated from the wear scar profile which unavoidably includes compacted oxides in case of tests conducted in air.

5.4.1. Unmodified Alloy

Figure 5.20 (a) and (b), and (c) and (d) shows optical and SEM micrographs of fretting scar on unmodified titanium alloy surface respectively.

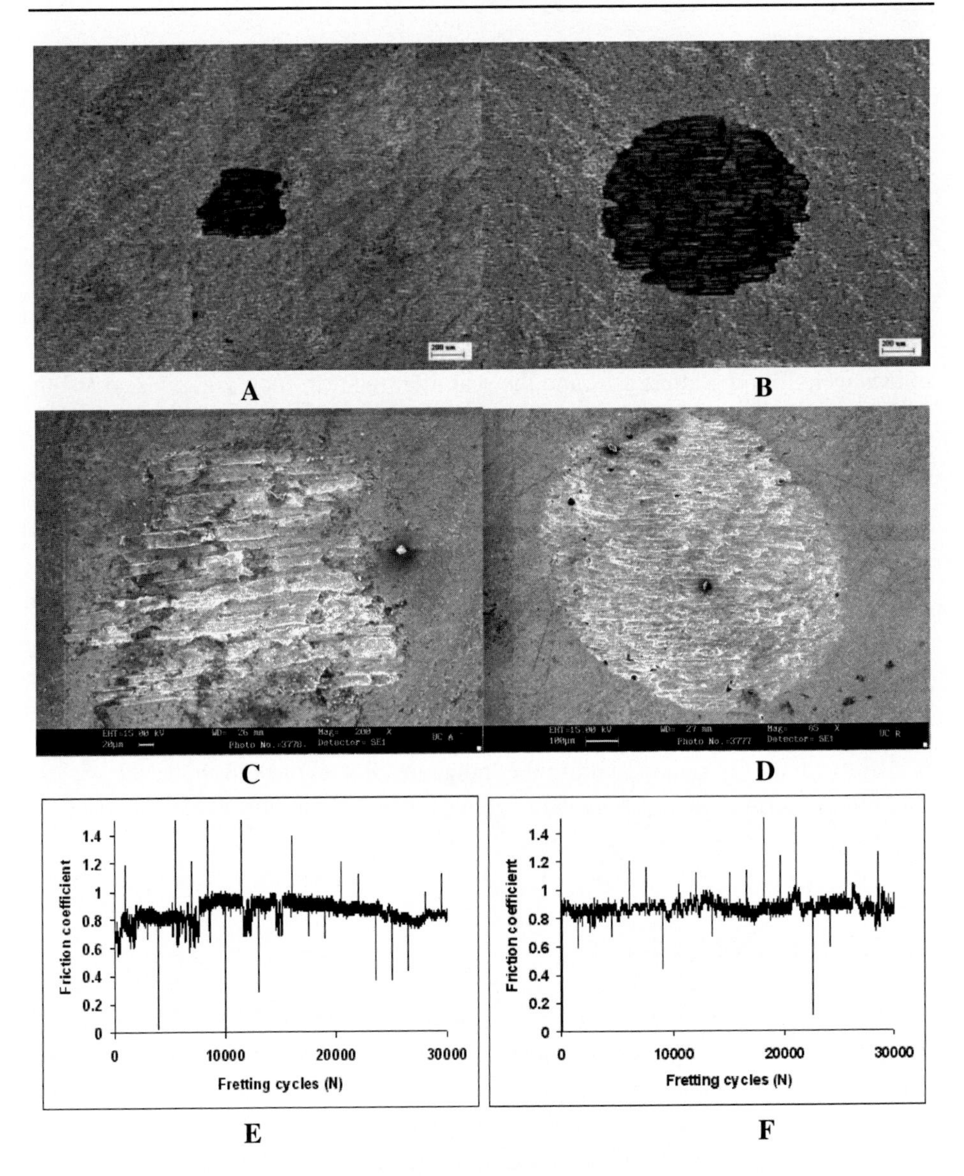

Figure 5.20. (a,c) Fretting scar of unmodified alloy in air, (b,d) Fretting scar in ringer fluid, (e) Friction coefficient curve in air, (f) Friction coefficient curve in ringer fluid.

Fretting scar is bigger and uniformly circular in ringer solution which is not observed in the fretting scar tested in air. Rutile is commonly formed product during fretting of titanium based alloys. The nature of debris formed very much influences the fretting behavior. In case of fretting in air, the oxides form at the contact and gets compacted within the scar. The transfer of products was also observed on the ball at contact point. Due to the presence of oxides, the abrasive counter body (alumina ball) cannot dig the surface effectively during fretting. The humidity of the atmosphere also influences fretting behavior (Chen et. al., 2002). Therefore the fretting scar width is less compared to that of ringer fluid in Figure 5.20 (b) and (d). The coefficient of friction is quite high compared to all other cases as will be seen later. In case of fretting in ringer fluid, the oxidized product is ejected out exposing fresh surface to mechanical attrition.

The compacted oxides are loosened by interaction with the fluid. Black oxides were found scattered around the scar after the test. Increasing in scar width is due to combined effect of mechanical attrition and dominating electrochemical oxidation (Gallino et. al, 2001).

The repeatedly exposed area at every fretting cycle acts anodic to the surrounding undamaged cathodic region for the electrochemical oxidation. The reacted area becomes weaker easily loosing more material during fretting even if the friction is lower. Wear particulates generated during fretting motion at the contact also take part in abrasive action (three body wear) of the layer along with electrochemical activity. Abrasive particulates can generate from both the modified layers as well as alumina ball. Figure 5.20 (e) and (f) shows friction coefficient curves for fretting wear of unmodified alloys in air and ringer fluid. It is observed to be steady within the range of 0.8 to 1.0 in both the cases. Intermittent formation and removal of oxides may be responsible for low friction events during fretting.

5.4.2. PVD TiN Coating

Figure 5.21 (a) and (b) and (c) and (d) shows optical and SEM micrographs of fretting scar on PVD TiN coated titanium alloy. The TiN layer is naturally more resistant to abrasion and electrochemical oxidation process compared to fretting of unmodified alloy. In case of air, fretting mark is little deep compared to ringer fluid. The oxides are also observed during test. Nano crystalline or amorphous rutile or anatase is a possible oxides produced from fretting on TiN layer (Chen et. al., 2005).

Figure 5.21 (e) and (f) shows friction coefficient curves for PVD TiN coated alloy. It is observed that friction is steady and lies below 0.2 in both air and ringer fluid throughout the test. This shows the superior tribological properties of TiN layer towards fretting wear resistance.

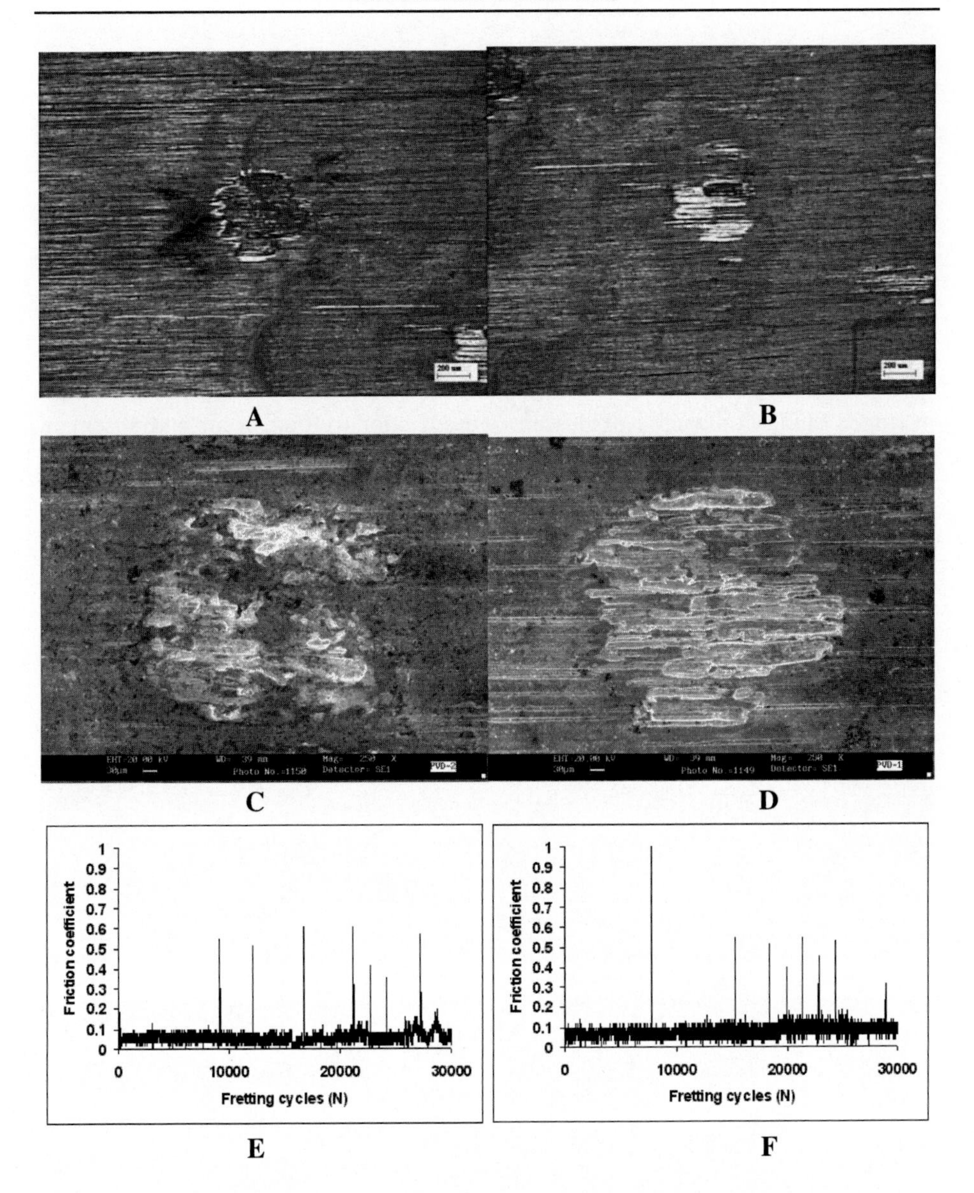

Figure 5.21. (a,c) Fretting scar of PVD TiN coated alloy in air, (b,d) Fretting scar in ringer fluid, (e) Friction coefficient curve in air, (f) Friction coefficient curve in ringer fluid.

5.4.3. Plasma Nitriding

Figure 5.22 (a) and (b), and (c) and (d) shows optical and SEM micrographs of fretting scar on plasma nitrided titanium alloy surface. A substantial difference in the average wear scar diameter is observed in ringer solution and air. The shows that it is difficult to penetrate the layer in air than ringer fluid. The worn area is also very irregular and rough with debris particulates. The nitride debris appears to dwell in the contact even during fretting. The layer removal is also not very uniform like that of Ringer fluid. The fretting scar profile is similar to unmodified alloys with oxidation in air and electrochemical oxidation phenomena in ringer fluid. Compacted oxides are present after fretting in air and absent in Ringer fluid. The surface roughness also plays a role during fretting. Smoother surface will have higher magnitude of real contact and should normally experience higher friction. Rougher surface will have to initially fret and stabilize the peaks and valleys before it can get the real contact area.

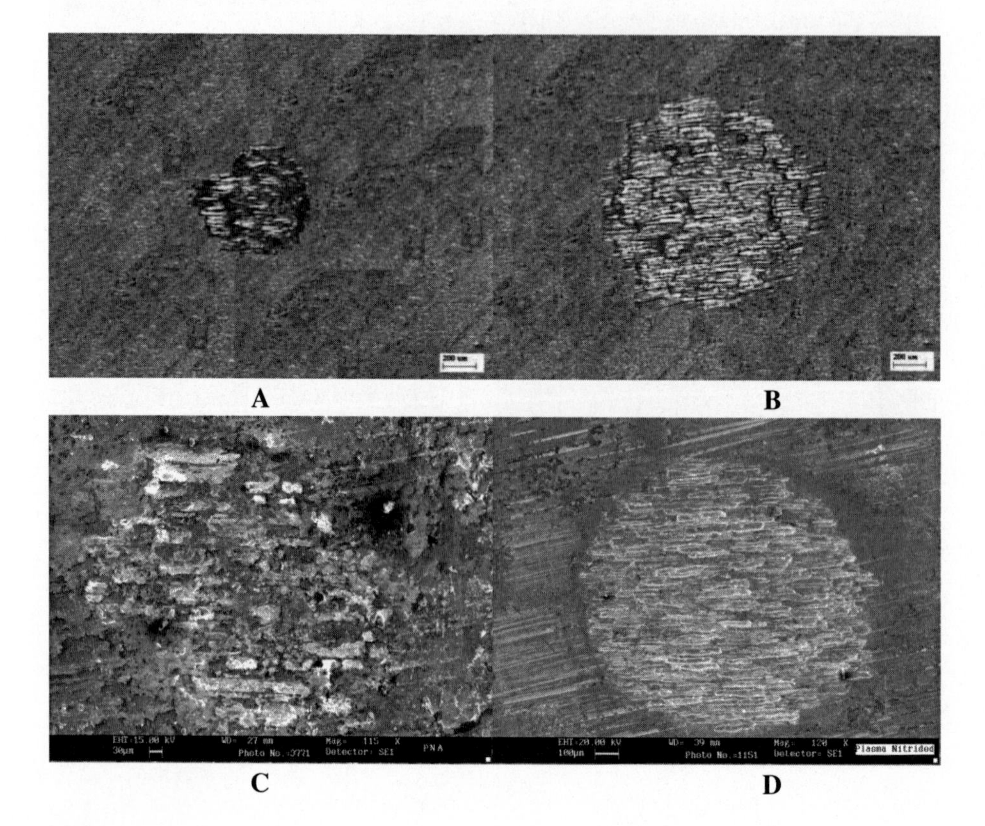

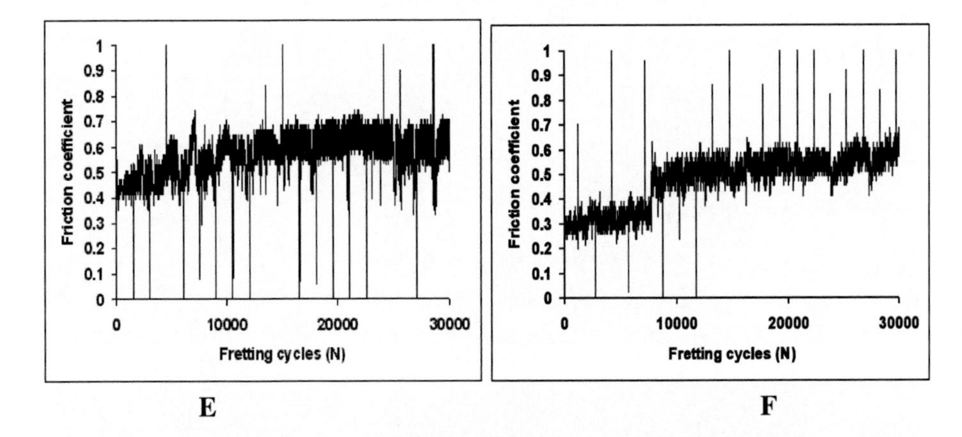

Figure 5.22. (a,c) Fretting scar of plasma nitrided alloy in air, (b,d) Fretting scar in ringer fluid, (e) Friction coefficient curve in air, (f) Friction coefficient curve in ringer fluid.

Figure 5.22 (e) and (f) shows friction coefficient curves in air and ringer fluid. In both the cases steady increase in friction from 0.4 to below 0.7 is observed. However there is an abrupt rise in friction in ringer fluid after 8000 cycles but below 0.6. This is possibly due to fretting on the substrate after the modified layer has eroded away.

5.4.4. Nitrogen Ion Implantation

Figure 5.23 (a) and (b) and (c) and (d) shows Optical and SEM micrographs of fretting scar on nitrogen ion implanted titanium alloy surface. The average wear scar diameter is almost similar in ringer solution as well as in air. However, the scar profile is more irregular in air and regular in Ringer fluid. Fretting scar for specimen tested in Ringer fluid has generated a larger and deeper scar due to fretting induced electrochemical dissolution process similar to unmodified and plasma nitrided alloys. Compacted oxides were observed within the fretting scar for specimen tested in air. Whereas the oxides were found scattered around the contact for specimen tested in Ringer solution. The implanted layer being softer and thinner could not protect the surface for a longer period.

Figure 5.23 (e) and (f) shows friction coefficient curves for fretting studies on ion implanted titanium alloy. It can be observed in Figure 5.23e that film has ruptured after 2500 cycles increasing the friction coefficient from 0.6 to 0.8 during fretting in air. Whereas in ringer fluid it has steadily maintained between 0.5 and 0.75 although it has penetrated the modified layer.

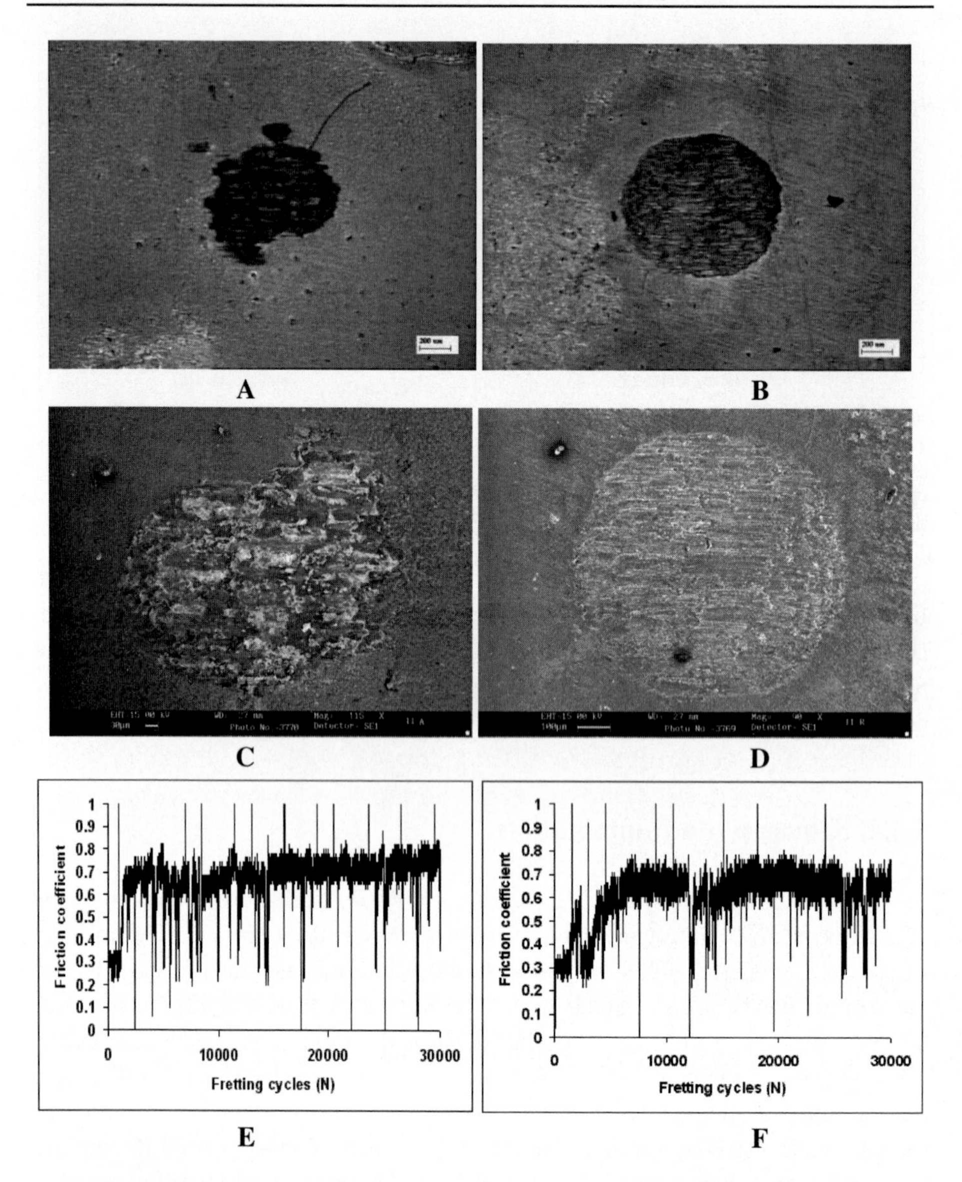

Figure 5.23. (a,c) Fretting scar of nitrogen ion implanted alloy in air, (b,d) Fretting scar in ringer fluid, (e) Friction coefficient curve in air, (f) Friction coefficient curve in ringer fluid.

5.4.5.Laser Nitriding

Figure 5.24 (a) and (b) shows optical micrographs of fretting scar on laser nitrided titanium alloy surface. The scar profile could not be observed in SEM due to shallow area. Average scar diameter in ringer solution is smaller compared to that of air. The damage is minimal in laser nitrided alloys compared to all other cases even though steady state friction coefficient is reasonably high compared to PVD TiN. It can be readily attributed to the presence of large volume fraction of closely spaced TiN dendrites in the modified volume as well as high roughness. TiN dendrite is harder compared to matrix.

Since it is present in substantially large volume percent, it acts to protect the bulk from higher rate of material loss. Dendritic TiN is highly hard and stable than the matrix surrounding it and high friction could be due to mechanical interlocking of projected dendrites and roughness peaks with the alumina ball surface during fretting.

Figure 5.24 (c) and (d) shows friction coefficient curves for fretting on laser nitrided alloy. It can be observed that in both cases, the friction coefficient is steady. It is almost 0.65 in case of air and 0.35 in case of fretting in ringer fluid with a slight decreasing trend during later stages. Compared to other processes it can be observed that the friction coefficient does not vary in wide range here. It is also noted that the surface damage profile is far less compared to other cases even though friction coefficient is higher than say PVD TiN coating. Therefore, there is a possibility that the material loss may have occurred to the counter body (alumina ball) than to the nitrided layer. The severity of the wear damage always depends upon the relative hardness. The harder material damages the softer one during wear process. Since the TiN dendrites are substantially harder than alumina counterpart is, the damage is less severe to the nitrided surface. The magnitude of friction also depends upon the roughness at the contact.

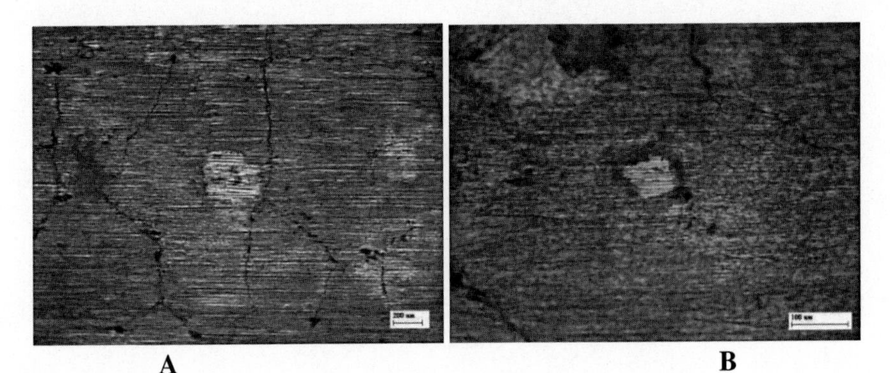

A **B**

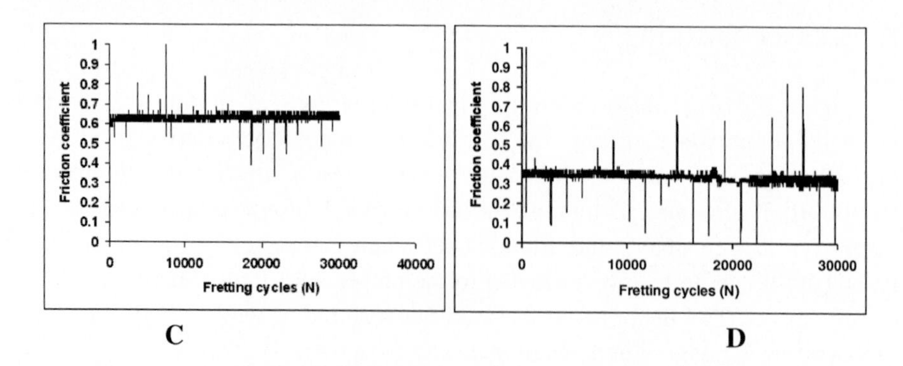

Figure 5.24. (a) Fretting scar of laser nitrided alloy in air, (b) Fretting scar in ringer fluid, (c) Friction coefficient curve in air, (d) Friction coefficient curve in ringer fluid.

5.4.6. Thermal Oxidation

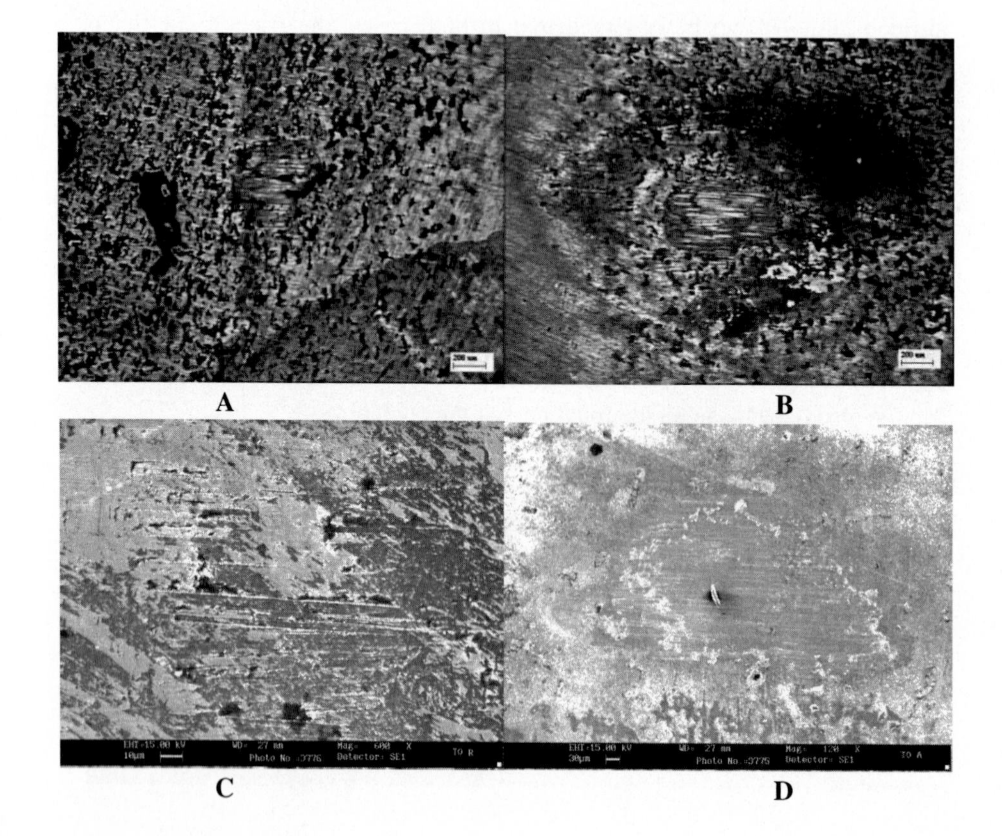

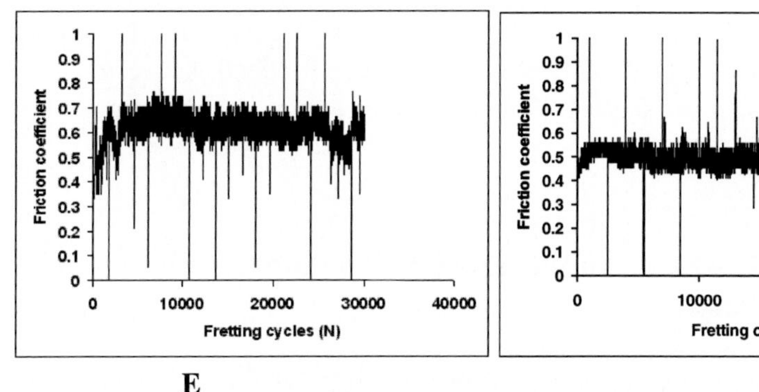

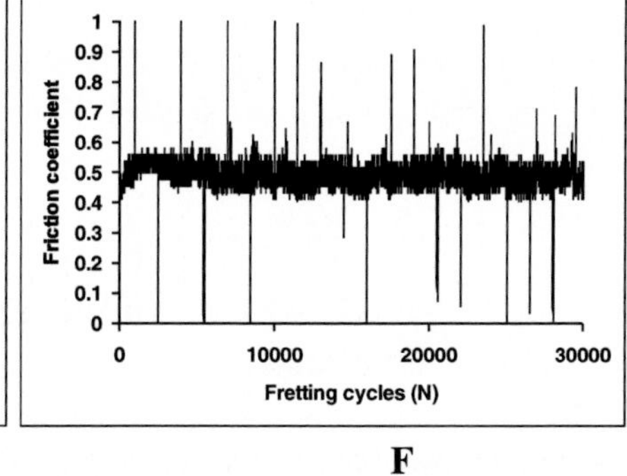

Figure 5.25. (a,c) Fretting scar of thermally oxidized in air, (b,d) Fretting scar in ringer fluid, (e) Friction coefficient curve in air, (f) Friction coefficient curve in ringer fluid.

Figure 5.25 (a) and (b) and (c) and (d) shows optical and SEM micrographs of fretting scar of thermally oxidized titanium alloy surface. The surface coating on PVD TiN is uniform all over the substrate and surface modification through ion implantation is also uniform over the entire implanted surface. Both processes do not allow any post surface preparations and the substrate has to be properly prepared prior to any modification.

However, in case of thermal oxidation, although the surface is uniformly oxidized through out the substrate material, the development of oxidized case is highly non uniform below the black rutile layer. Since the interface between the rutile and oxidized α case is highly irregular or wavy by nature, it poses difficulty in producing smooth surface for fretting tests. The nature of the surface is shown in Figure. 5.25 (a) and (b). Some amount of modified case is also removed during surface preparation through polishing. This indicates that the oxygen diffusion into the alloy is not very uniform during the process of oxidation. Polishing process has removed the case completely in some region leaving the surface with varied distribution of α islands. Fretting test is therefore performed on a selected area where there is a high probability of undisturbed α case. Even though fretting scar width is less both in air and in Ringer fluid, the oxidized alloy has produced high friction coefficient throughout the test due to irregular surface profile.

The average wear scar diameter is smaller in air compared to ringer solution. The scar profile is irregular in both cases. The oxidized case, even though irregular, is thick in addition, hard enough for preventing deeper damage during fretting compared to other processes. The wear tracks are seen aligned along the fretting direction.

Figure 5.25 (e) and (f) shows friction coefficient curves for fretting on thermally oxidized titanium alloy. Here the steady state friction is between 0.6 and 0.7 after gradual increase from 0.4. Whereas in ringer fluid, it ranges from 0.4 to 0.6.

5.4.7. Comparative Study of Steady State Friction Coefficient and Wear Scar Diameters

Figure 5.26(a) shows the comparison bar chart of steady state friction coefficient for all the surface treated alloys tested for fretting wear in air and ringer fluid. Friction coefficient is higher in air than ringer solution for all the cases, except PVD TiN coating which could be due to dominating asperities, oxides and wear debris interaction at the contact. Ringer solution also offers lubricating effect during fretting reducing the friction at the contact in all other cases. PVD TiN coating has shown the lowest friction coefficient (10 – 20%) compared to all other coatings.

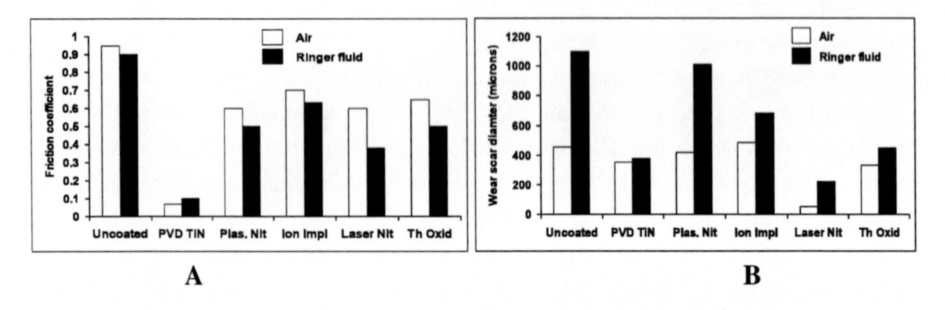

Figure 5.26. Comparison chart of (a) Steady state friction coefficient, (b) Wear scar diameter for fretting wear tests.

Figure 5.26(b) shows comparison of average fretting wear scar diameter for all the cases. The scar profile and friction depends very much on (a) Layer hardness, (b) Layer thickness, (c) Layer microstructure, (d) Fretting medium and (e) Products formed within the scar during fretting. The wear scar generated during fretting is irregular shaped in air and regular circle with definite diameter in ringer fluid in unmodified, plasma nitrided and ion implanted samples. The wear scar diameters in ringer fluid is larger compared to air in all the cases. Large differences in wear scar diameters are observed in case of laser nitrided, unmodified and plasma nitrided alloys. Finally PVD TiN coating and laser nitriding proves to give best performance in terms of minimal fretting damage and

PVD TiN gives minimum damage as well as reduced coefficient of friction for 30,000 fretting cycles. Thermal oxidation has given intermediate performance. The wear tracks developed due to fretting is shown at higher magnification SEM micrograph in Figure 5.27. Tracks are all aligned parallel to fretting direction. The same is observed in case of unmodified, plasma nitrided and ion implanted alloys tested in ringer solution. PVD TiN coating showed good fretting resistance both in air as well as ringer fluid.

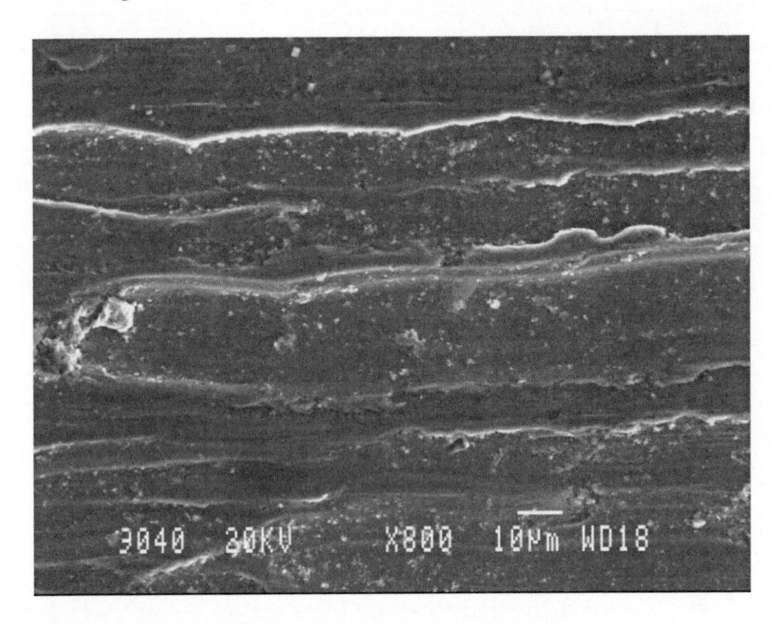

Figure 5.27. Wear tracks within the fretting scar of unmodified alloy.

5.4.8. Fretting Wear Scar Depth and Wear Rate Comparison

Figureure 5.28 (a) and (b) shows fretting wear scar depth comparison in air and ringer solution. It is clearly evident that laser nitriding and PVD TiN coatings have shown good performance over all other coatings with minimum surface damage. Laser nitrided coatings have proved superior even to PVD TiN coating. The scar width is the smallest in air for laser nitrided sample as seen earlier and the depth is also very shallow. The scars were difficult to locate even with naked eye in both these cases. The scar is broad and shallow in both cases. The scar depth is higher for softer thin film layers and unmodified samples. The magnitude of scar depth is intermediate for thermally oxidized samples. The depth is irregular in all the cases because of mechanical abrasion in two body and three body wear mode. In ringer fluid it is little more uniform due to influence of electrochemical activity during fretting. It is also observed that depth is more in ringer fluid for unmodified, plasma nitrided and ion implanted alloys than other coatings. This indicates that the base material is very much influenced by the medium during fretting. The influence of corrosion during fretting although not studied here is expected to have occurred favoring easier removal of material.

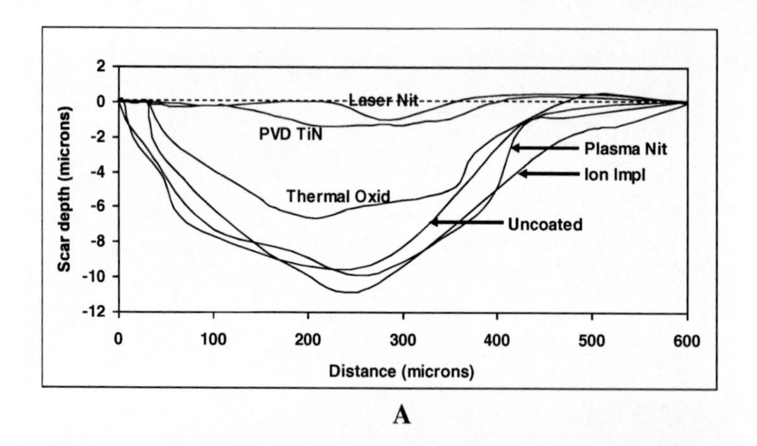

A

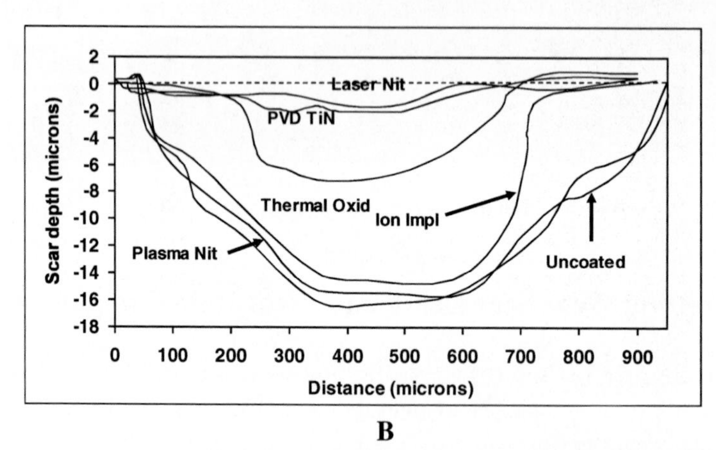

B

Figure 5.28. Fretting wear scar depth profiles in (a) Air and (b) Ringer solution.

Fig 5.29 shows comparison chart of wear rate during fretting for all the surface modified layers. It is clearly evident that PVD TiN coatings, laser nitriding and thermal oxidation have shown low wear rates compared to other layers. The wear depth is very shallow for these layers and hence volume of material removed is naturally less compared to others. It can also be inferred that the wear damage is much more reduced for laser nitrided case compared to PVD TiN even though the laser nitrided layer has experienced higher friction coefficient during the fretting wear test.

This indicates that TiN developed through laser nitriding is much more superior compared to PVD process. PVD process develops TiN in columnar structure with high residual stresses between the columns as explained earlier in section 5.3.1. Whereas the laser nitriding process produces TiN in dendritic form during melting and solidification of the substrate. This form of TiN has shown a slightly superior wear resistance compared to PVD TiN layer. It may be related to fretting induced relaxation of residual stresses developed by PVD coating process. This can finally break the TiN columns perpendicular to the growth direction.

Wear rate is higher in ringer fluid than air in all cases. The difference is especially large in case of unmodified, ion implanted and thermally oxidized specimens. The wear rates for laser nitrided, PVD TiN coated and thermally oxidized samples is 0.6, 3 and 12.5% respectively compared to other three layers.

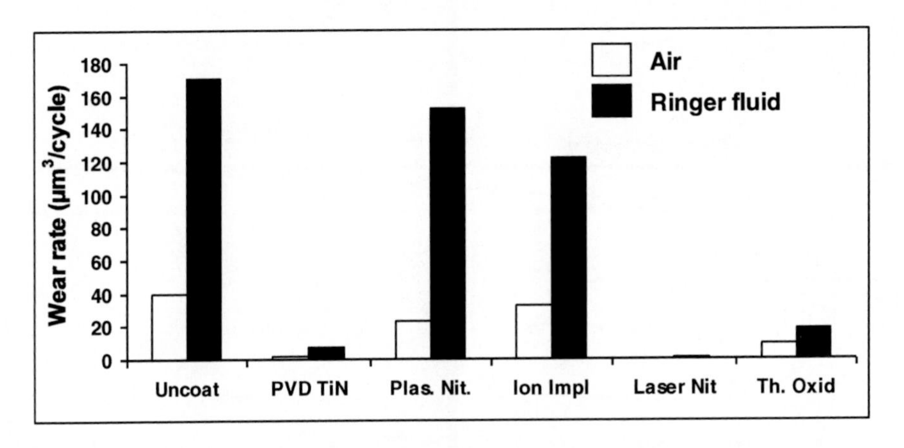

Figureure 5.29. Wear rate comparison for fretting tests of all the surface modified alloys.

5.4.9. Comparison of Fretting Loops

Figure 5.30 shows fretting hysteresis loops generated for different coatings after 30,000 fretting cycles in air and ringer fluid for 1N normal load. The area of hysteresis loop generally describes about the frictional energy dissipated during process of fretting.

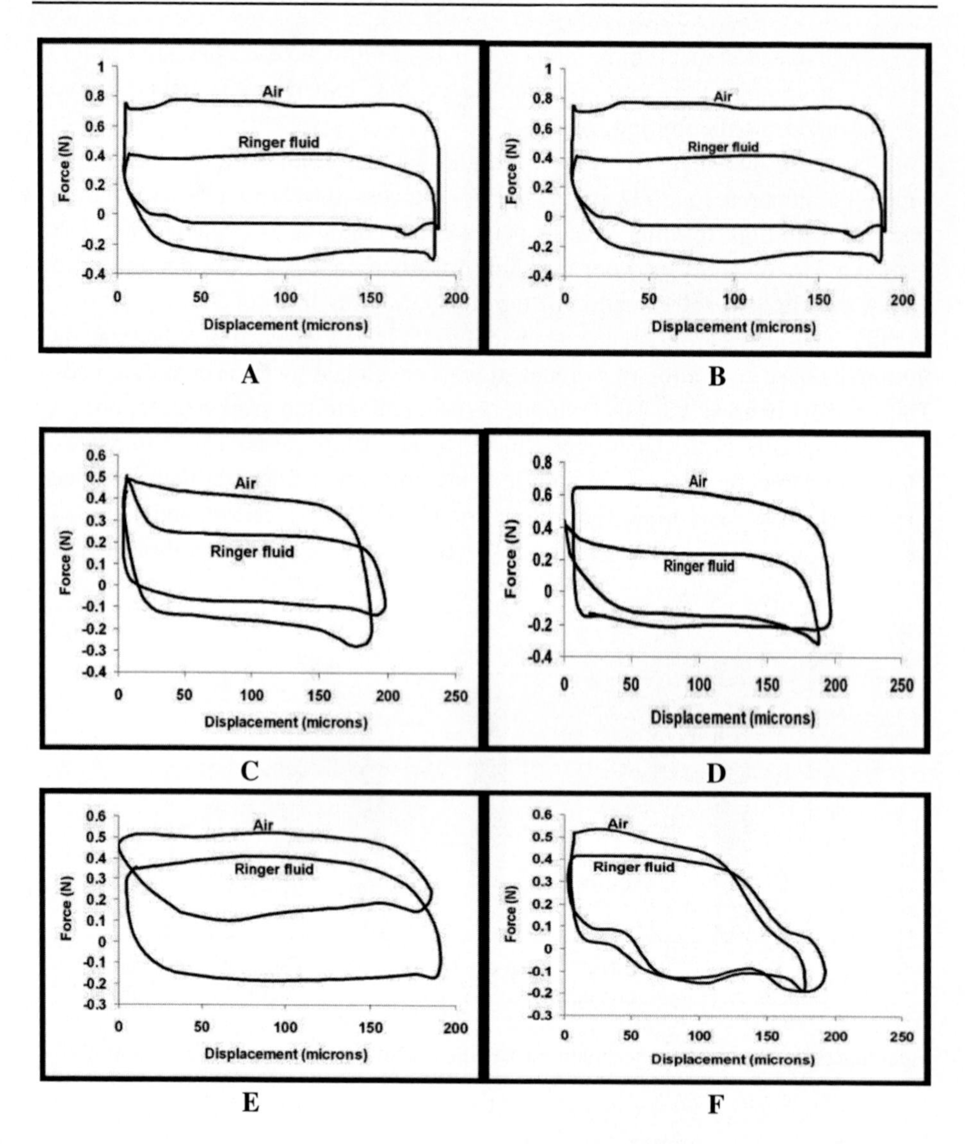

Figure 5.30. Force – displacement (F-D) loops after 30, 000 fretting cycles for 1N normal load in Air and Ringer fluid (a) Unmodified, (b) PVD TiN coated, (c) Plasma nitrided, (d) Nitrogen ion implanted, (e) Laser nitriding, (f) Thermal oxidation.

In general, larger area of the loop indicates more energy dissipation in the form of friction at the contact. In every case there is a difference between loops generated in air and ringer fluid. The loop pattern will be almost similar to a rectangle in the beginning. The changes in contour occurs only when the friction is altered by various parameters such as relative humidity in air, ringer fluid and debris formation.

Oxide debris plays a major role during the process. The contour of the hysteresis loop is almost unaltered in case of unmodified alloys and PVD TiN coated alloys in air and ringer fluid.

Figure 5.31 shows force – displacement (F-D) loops for fretting of unmodified Ti-6Al-4V and hardened steel ball configuration at different loads. We can observe that loop pattern changes with increasing normal load. It changes from rectangular shape at lower loads to oval shape at higher loads. This indicates the shift in the fretting regime, i.e. changing from gross slip to partial slip regime. The frictional energy dissipation reduces with the change in fretting regime. At lower loads, surface asperities plays role in altering the friction. At higher loads, surface and subsurface plastic deformation dominates more than the asperities interactions. Usually friction coefficient reduces with increasing normal load due to plastic flow at the contact as seen from Figure 5.32.

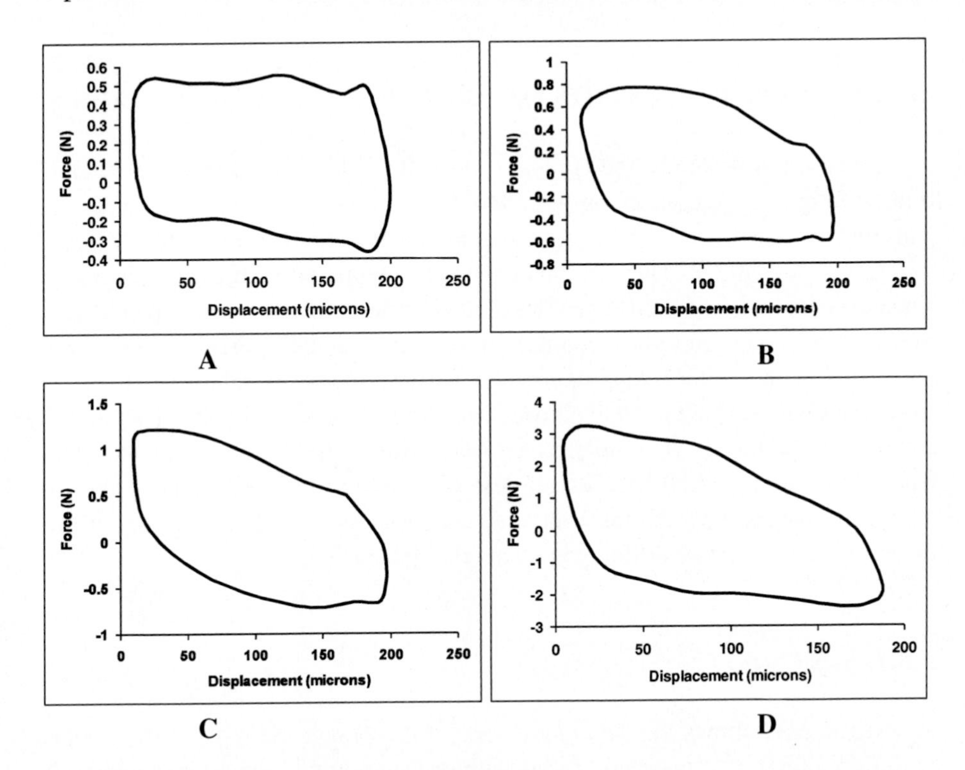

Figure 5.31. Force – displacement (F-D) loops for different normal loads with hardened steel ball on unmodified Ti-6Al-4V alloy. Normal Load: (a) 1N, (b) 5N, (c) 8N, (d) 25N

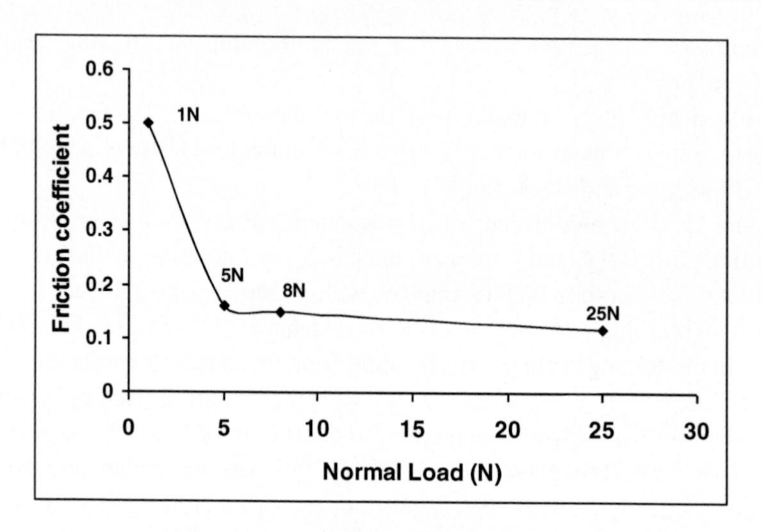

Figure 5.32. Friction coefficient for different normal loads.

5.5. Characterization of Fretting Fatigue Failures

This section describes the comparison of lives of modified specimens under fretting fatigue conditions along with variation in relative slip range between the pads and the specimens. This is followed by detailed analysis of surface damage during fretting fatigue. These results cannot be compared with the fretting wear since the counter body is different. The contact behavior very much depends upon the counter body and the medium around the contact. Fretting wear was conducted with alumina ball on all the surface modified samples. However, fretting fatigue tests are conducted with two similarly modified pairs. Therefore it cannot be correlated with the fretting wear results. In fretting wear tests, the hardness of alumina ball is relatively less compared to unmodified alloy. So the damage is substantially severe. Where as the hardness of PVD TiN film and TiN developed through laser nitriding is relatively higher.

5.5.1. S-N Curve

Figure 5.33 shows the S-N curve for fretting fatigue tests of all the surface modified alloys. It can be compared with the unmodified specimens and plain fatigue tests. The tests for all conditions were conducted until the ultimate failure of the specimen. A drastic reduction in lives of both surface modified and unmodified alloys has been observed compared to plain fatigue lives. The arrow indicates the specimens which did not fail after a million cycles. At 500 MPa cyclic stresses, the fretting fatigue life of unmodified specimen is just 3% of the plain fatigue life as shown in Figure 5.33. This highlights the conspicuous effects of fretting on the life of titanium alloys.

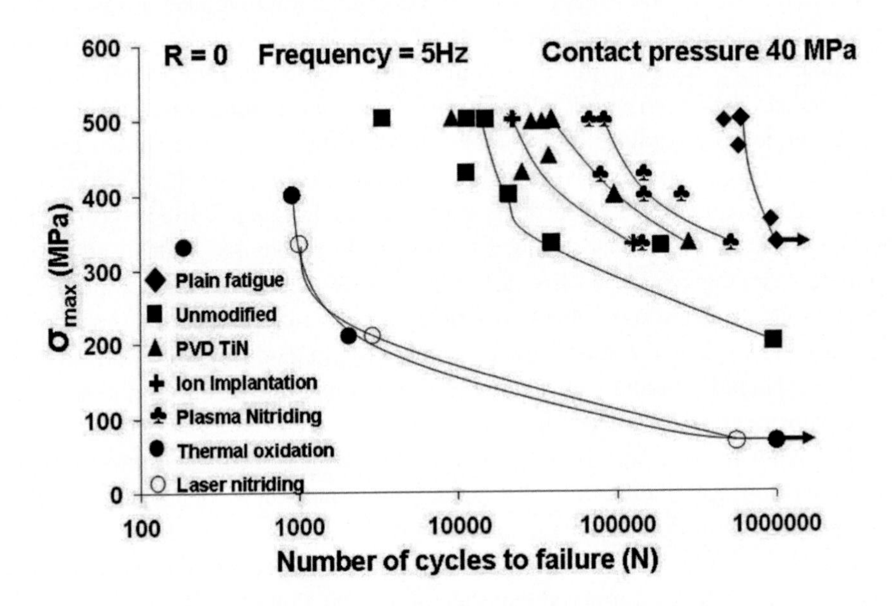

Figure 5.33. S-N curve for titanium alloys under different conditions.

Plasma nitrided pairs have shown the best performance over all the coatings due to gradient modified layer which have high strain accommodation capacity during fretting induced surface and subsurface plastic deformation compared to other layers. For example, the fretting fatigue life of unmodified alloy pair is just 15 to 18% of the life of plasma nitrided pair at 500 MPa cyclic stresses. This shows the effect of modified layer on fretting fatigue strength. Although the final surface damage is similar to unmodified alloys, the layer is more effective during the initial period. PVD coatings failed earlier compared to plasma nitrided pairs possibly due to severe third body wear mode of contact interaction in the later stages of fretting. But life is improved compared to unmodified alloys. For example, the fretting fatigue life of unmodified alloy is 45 to 50% of the PVD TiN coated alloy at 500 MPa cyclic stresses. Whereas the life of PVD TiN coated couple is almost 50% of the plasma nitrided couple at 500 MPa cyclic stresses. The improvement is naturally much larger at lower stresses as seen from the S-N curve. Plain fatigue life of PVD TiN coated alloy crossed one million cycles without failure. It is due to the effect of hard layer on the surface preventing the formation of extrusions and intrusions which acts like micro notches for initiating fatigue cracks.

Ion implantation showed slight improvement (20 to 25%) over the unmodified alloys at 500 MPa axial stresses and good improvement at lower axial stress. Due to complexity of the treatment and process constraints, only two specimens were tested with ion implantation. Laser nitriding and thermal oxidation has shown very poor performance. At higher nominal stress levels, they both experienced premature failure and at lower stress, they have shown infinite lives with intangible fretting at the contact. This is possibly from the effect of high case depth in case of laser nitrided samples and irregular surface profile for thermally oxidized specimens. When tensile load exceeds the yield limit of the modified case, the specimen fails earlier. In case of thermally oxidized pairs, the specimen failure is primarily due to irregular case development during the thermal processing as explained earlier. The region between unmodified and oxidized case acts like a potential failure site. This will be explained further during the analysis of the damage profiles for each of the above cases.

The main interest here is only to show the improvement of fretting fatigue lives due to surface modification on titanium alloys and not to identify or optimize any specific parameters. Average life of any specimens under fretting fatigue conditions can only be adjudged after testing many specimens in similar load conditions. Usually it is observed that there would be a lot of scatter in the lives of specimens under plain fatigue conditions itself due to inevitable differences in test specimens.

Recent trend is to obtain the Gaussian distribution of lives for a single cyclic load rather than plotting an S-N curve. Many specimens are tested for same load and a record on distribution of lives would give us an apparent idea of the fatigue life of the material of a particular geometry. In fretting fatigue experiments it is more difficult to get even a rough estimate of life at a particular load history due to more complexity in stresses involved at the contact area. Therefore there are no standards followed in fretting fatigue experiments as of today.

The tendency for crack initiation also depends upon the surface microstructural homogeneity or uniformity, which may vary for different specimens under test. The compacted debris in itself may influence the severity of fretting. Harder transformed products may be severely damaging whereas softer products may lubricate the surface and reduce friction.

Figure 5.34 shows the average roughness at the contact area before and after fretting fatigue test. The damage is severe for unmodified, PVD TiN, plasma nitrided and ion implanted alloys. It is to be expected because all these coatings have endured longer fretting fatigue cycles with gradual increment of damage. The roughness value also includes that of compacted rutile oxides within the damaged area, as it could not be removed. Laser nitrided and thermally oxidized pairs have not shown any marked changes due to premature failure of the samples as seen from the S-N plot.

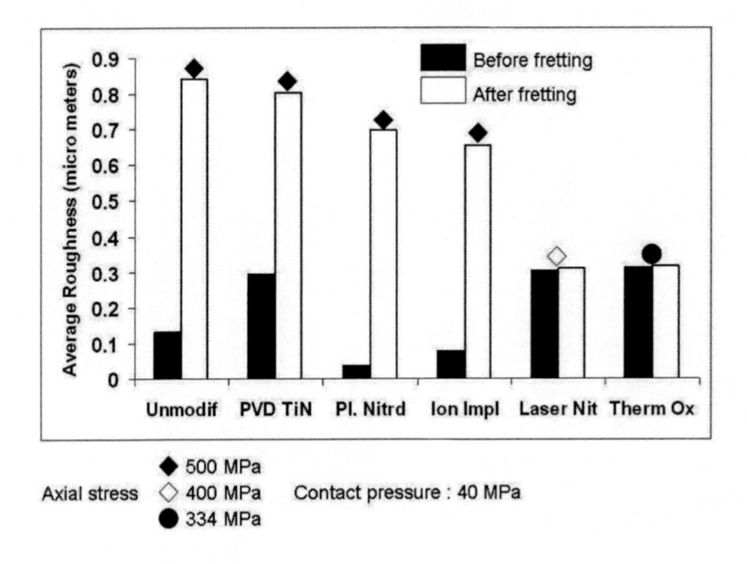

Figure 5.34. Comparison of roughness before and after fretting fatigue tests.

5.5.2. Slip Amplitude

Figures 5.35 (a) and (b) shows the variation in slip range with the applied static axial stress in air and Ringer fluid. It can be observed that slip increases with increase in maximum applied stress in air as well as ringer fluid. Slip experienced by contact pairs in ringer fluid is slightly higher compared to air due lubricating effect. They appear almost equal at same applied stresses because the asperities interlocking during sliding is released by flow of ringer fluid. The effect of asperities is more observed in air than ringer fluid. Surface modified pairs have shown slightly higher slip compared to unmodified pairs in air as well as fluid. Even unmodified pairs have thin adherent oxide layer which will eventually break during cyclic fretting. When the friction between mating pairs is less, the relative sliding is higher and vice versa.

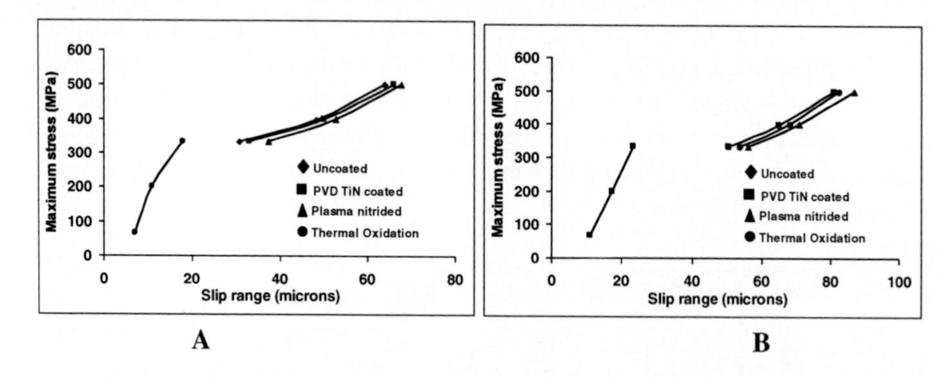

Figure 5.35. Variation in slip range with maximum stresses under static loading conditions in (a) Air, (b) Ringer fluid.

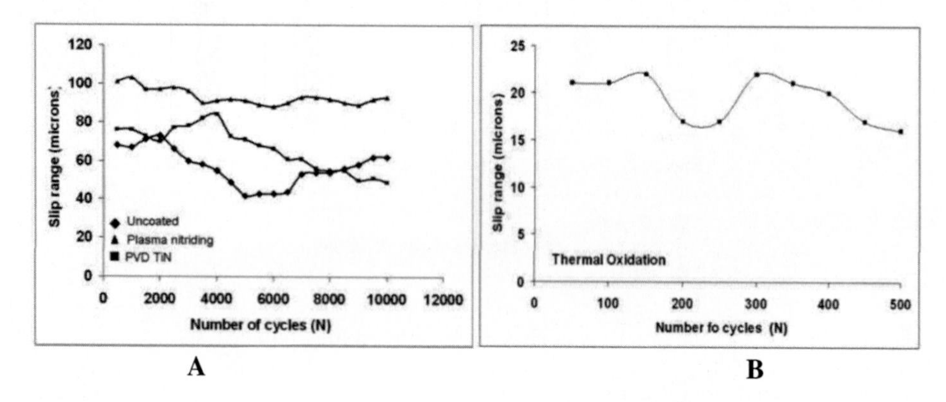

Figure 5.36. Slip range variation during the progression of fretting fatigue.

Figure 5.36 (a) and (b) shows variation in slip range with fretting cycles. The test is conducted only for 10,000 cycles. As shown in Figure 5.36a, larger variation is observed in case of unmodified and PVD TiN coated pairs. Where as the slip range variation is almost consistent with plasma nitrided pairs. Fretting of unmodified pairs shows high metallurgical compatibility as said earlier undergoing frequent adhesion or micro-welding of projected asperities. Adhesion and break away of asperities affects the slip regime during fretting fatigue as observed in the figures. In case of PVD TiN coated alloys, the process of adhesion is delayed for some period and the irregularities in slip amplitude are mainly due to particulate production and entrapment during the course of fretting. However, in both cases, higher slip amplitude indicates low friction between mating pairs and vice versa. The variation in slip range of thermally oxidized alloy pair is shown in Figure 5.36b. The profile is similar to former case but with reduced slip range values.

5.5.3. Fretting Fatigue Surface Damage Characterization

In this section, the fretting fatigue damage features is explained for each individual case beginning with unmodified pairs. The damage features were almost similar at all the contact points except in some cases. The damage progression during fretting process is explained with friction coefficient curves. It is assumed that friction is also affected by pad geometry apart from fretting during the test. Bridge type pad configuration offers advantage of measuring friction but also has disadvantage of contact gapping normally difficult to observe during test. The friction curve in each case here is the average of two pads on either side of the specimen. Therefore two curves are observed in each case. Surface damage micrographs shown is for 500 MPa axial stresses in case of unmodified, plasma nitrided, PVD TiN coated and ion implanted specimens and 400 MPa and 334 MPa for laser nitrided and thermally oxidized pairs respectively. All other conditions remain same for all the cases. EDS analysis shows the chemical composition at the damage site.

5.5.3.1 Unmodified Alloys

Fretting fatigue surface damage of unmodified alloy at 500 MPa axial loads is as shown in Figure 5.37 (a and b). It is characterized by heavily worn out surface in the form of deep scratches, particle detachment, oxidation and transfer from counter surface. Figure 5.38 is an EDS of the scar showing Nb peak which indicates the transfer of particles from the opposing pad surface. Both the specimen-pad couple, being the same material, exhibit high metallurgical compatibility which is the primary reason for such an extreme form of damage. Higher compatibility induces localized adhesion and mutual dissolution of the contact materials during fretting process. Frequent stick-slip at contact also produces squeaking sound which was intermittently witnessed during the test. The laminates and fine particles with high surface area are more prone to chemical reactions with tissues and protein which may induce allergic response. Figure 5.39 shows the roughness profile of the damaged area indicating extreme unevenness of the surface ($R_a = 0.84$ μm compared to $R_a = 0.11$ μm for as polished surface).

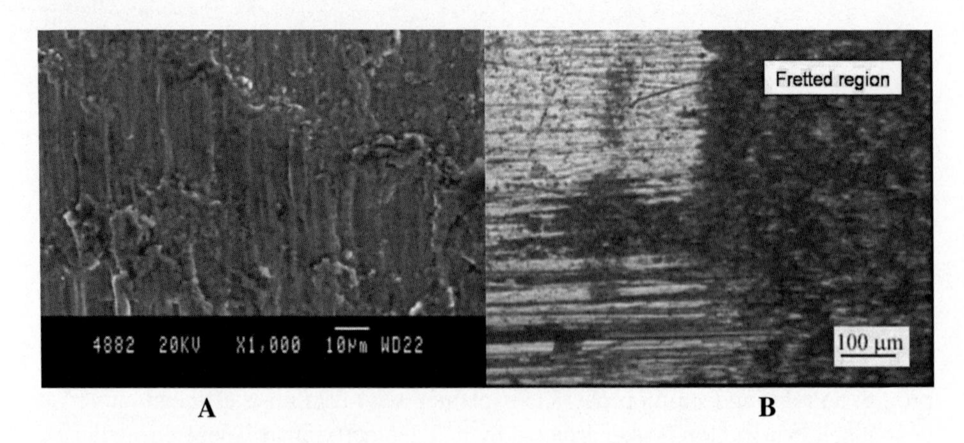

Figure 5.37. (a) SEM micrograph (b) Optical micrograph of fretting scar of unmodified alloys.

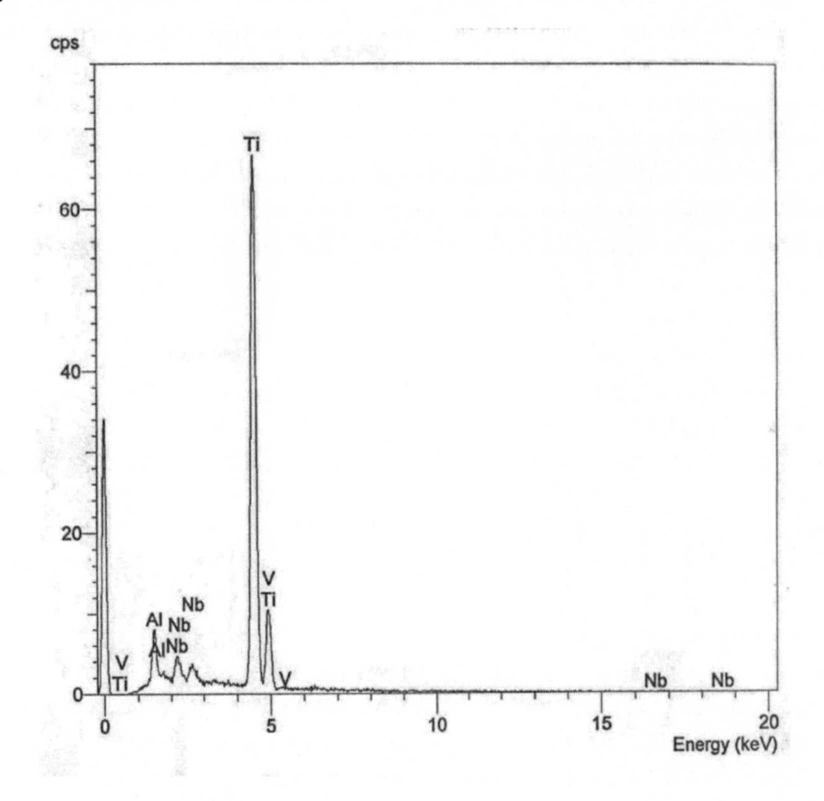

Figure 5.38. EDS of fretting scar of unmodified Ti-6Al-4V.

Figure 5.40 (a) and (b) shows the friction curves for the fretting fatigue tests of unmodified alloys at 500 MPa axial load. It is interesting to note that both the pairs of pads have experienced similar variation of friction during fretting. At higher axial load, this may be possible due to higher sliding amplitude when local surface inhomogenities are overruled. As said earlier, frequent localized-welding and rupture of junctions occurs during fretting of bare surface. When slip amplitude increases with increasing axial load, the rupture of localized welded junctions takes place more rapidly. Friction coefficient for 500 MPa varies from 0.3 to 0.6 and reaches 0.8 just before ultimate failure. It is also observed to steadily glide down after initial few cycles. This can be attributed to the combined lubricating effect of biofluid and oxide bed. The steady formation and ejection of oxides at the contact also minimizes the resistance to sliding motion. The degree of damage gradually compounds from frequent ejection of oxides with the flow of fluid and formation of fresh oxides from exposure of fresh surface. The saline fluid also mixes with the oxides which get compacted within surface cracks causing further dissolution of material at the crack tip. Higher friction values before ultimate failure indicates severity of damage. As seen earlier from the roughness profile, the pads would plough into the surface resulting in deep scratches and immediate crack initiation and propagation to failure within few cycles.

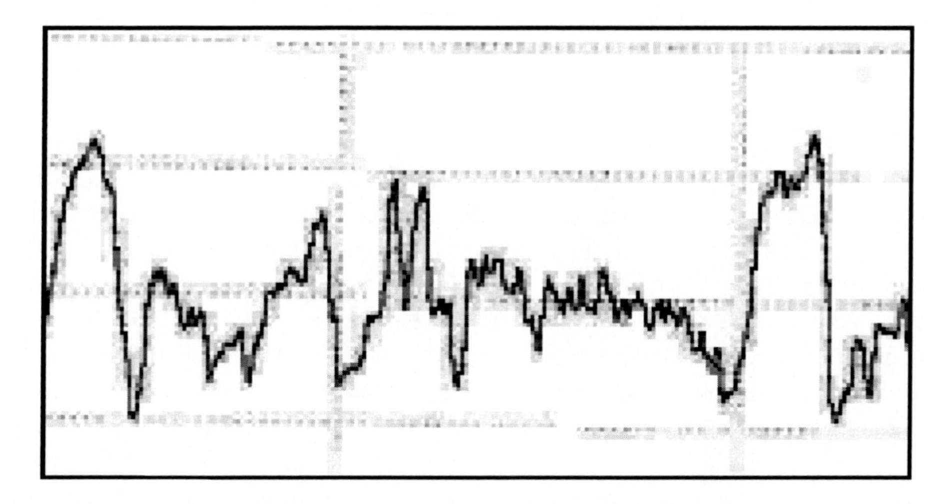

Figure 5.39. Roughness profile of the damaged site of unmodified Ti-6Al-4V.

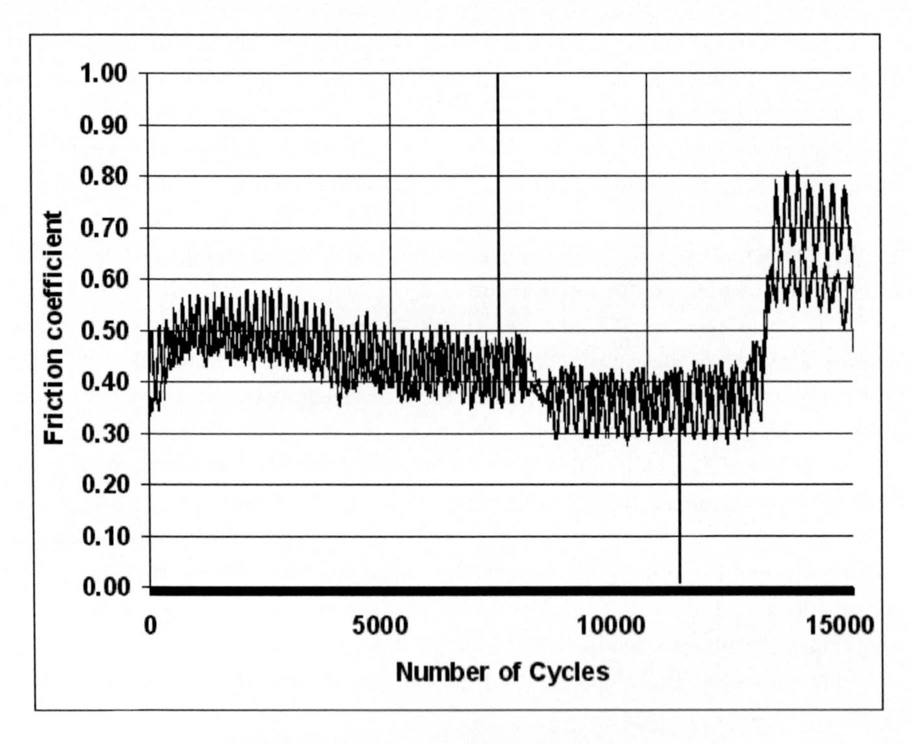

Figure 5.40. Friction coefficient curves of unmodified alloy pairs for cyclic stresses of 500 MPa.

5.5.3.2. PVD TiN Coating

For all the PVD TiN coated alloys, fretting cracks was observed to nucleate along the contact edge as well as within the contact region. Friction is minimized with TiN coatings due to its excellent tribological properties compared to unmodified alloys. But it could not withstand the fretting process for a long time when compared to unmodified alloys as observed in Figure 5.41(a). The lives of unmodified specimens were almost 45 to 50% of the PVD TiN coated samples. The film ruptured at the later stages of fretting resulting in gradually detachment of hard TiN layers. The ultimate rupture is preceded by the formation of network of cracks within the contact. Even the cracked layers can protect the surface for certain period of time until they are fragmented further. Some of the laminates are found deeply embedded within the underlying substrate material. Fretting motion under contact pressure has disintegrated the coated layer from the substrate and buried them deeper. The adhesion strength of the film depends upon the prior cleaning of the substrate and the purity of the film. Figure 5.41(b) shows the fractured cross sectional features of PVD TiN coated alloy. Fractured surface layer is not distinctly visible although some features of brittle rupture can be observed at the edge.

In these experiments, TiN layer is coated on both the specimen and pads. Since the hardness is the same, reasonable wear rate can be expected during gross sliding. It is also observed that some of the ejected TiN laminates had fretting marks on their surface. Surface projections initially rub against each other and ultimately rupture leaving these marks on the delaminated particulates. Uneven surface are sometimes beneficial in fretting application because the surface depressions can provide sink for the debris particulates preventing harmful effects.

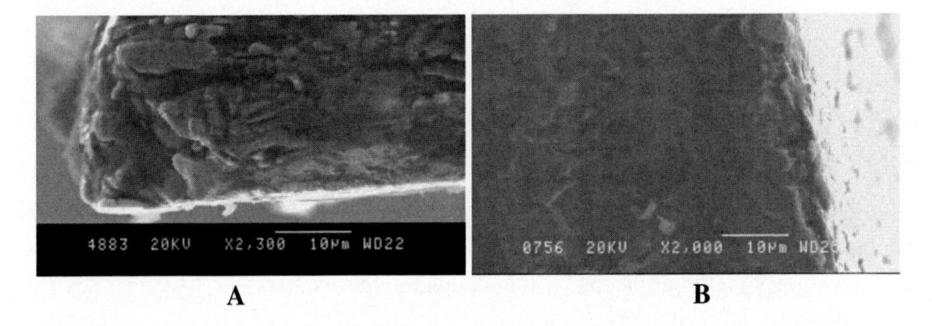

Figure 5.41. (a) Edge of contact failure area of PVD TiN coated alloy, (b) Cross section of fractured specimen.

Figure 5.42 shows EDS of ruptured area indicating Al and V peak from the exposed substrate. As mentioned earlier, these elements (even 10 ppb) are known to cause toxic effect and it is therefore important to maintain the continuity of the film for long term application of prosthetic devices. Delaminated TiN particles acted as potential abrasives for further damage on both the surfaces. Their sizes vary from less than a micron to more than 10 microns. Larger particles with sharp edges can be more harmful as they could easily plow into the surface during fretting. The stiffness of the film usually increases with increase in its thickness due to good packing and almost reaches theoretical density at higher thickness (Takahito Ohmura et al., 2003). When the flexibility is lost, it is easier to break the film by deflection or fretting. Some sort of surface preparation before coating may also help in altering the film properties because the substrate also plays a role in altering the nature of the film growth. The purity of the film also helps in retaining the continuity during sliding motion. Post coating heat treatment may help us to recover from some of these defects (Wen-Jun Chou et al., 2003).

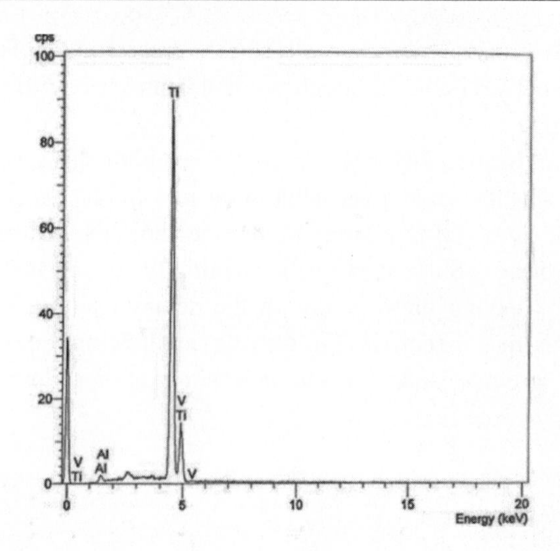

Figure 5.42. EDS of the damaged site of PVD TiN coated alloy indicating Al and V exposed from the substrate.

Figure 5.43 shows a crack on the TiN layer of a specimen perpendicular to the loading axis. Many such cracks are sometimes observed on the surface of the TiN layer especially adjacent to the contact edge. When the load crosses the elastic limit of TiN, such cracks appears distinctly all on surface because the layer cannot accommodate higher levels of surface strains like substrates due to high hardness (~40 GPa) and stiffness (~332 GPa).

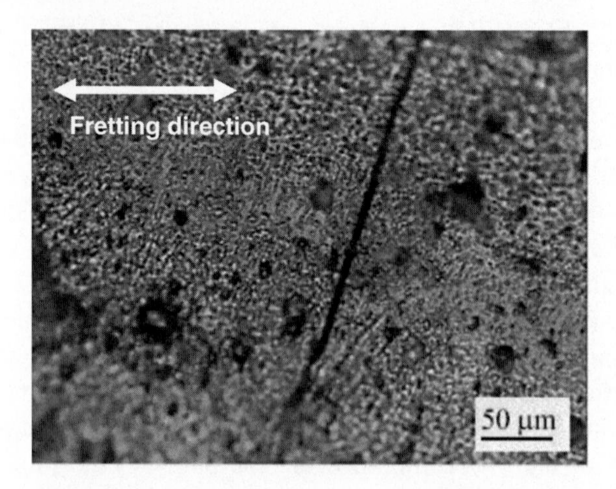

Figure 5.43. Cracks on the surface of PVD TiN coated fatigue specimen.

The substrate hardness is 4.2 GPa and stiffness is 104 GPa. Deformation is in plastic mode for substrate and elastic mode for the TiN layer. The axial load level was within the yield stress for the substrate but beyond the yield strength for the TiN layer. Therefore the TiN layer fails before the specimen. Figure 5.44 shows oxidation within one of the localized damage site. The weaker site on the film will expose the underlying substrate for oxidation when it ruptures.

Figure 5.44. Oxide within one of the localized damage site of fretting fatigue damaged specimen.

Figure 5.45(a) shows the friction curve for PVD TiN coated pair for 500 MPa axial loads. Friction coefficient varies from 0.1 to 0.35 throughout the test. Fretting process on TiN layer differ compared to unmodified alloy. So it is observed that friction is maintained within the limits till complete failure.

It is more unsteady compared to fretting of unmodified alloys as evident from the zigzag nature of the curve. Initial increase in friction is due to interaction of roughness peaks and later it goes down due to rupture of these projections. These ruptured particles forms initial debris bed within the contact and cause further release of particles from third body wear mode of contact interaction. After 20,000 cycles it is seen that there is a sudden increase in friction which can be attributed to delamination process and formation of oxide later reduces friction by fluid interaction and ejection of particles.

The formation of oxide always causes gradual or immediate slump in the friction due to smoother accommodation of tangential motion. Finer particles may or may not loose its effect from the compaction within the oxide debris. However it is not always applicable at all points of contact because each point experiences difference in contact behavior. These curves are the average values of two contact pads sliding on either side of the specimen.

Figure 5.45(b) shows the friction curve for 5000 cycles at reduced contact pressure of 20 MPa. The unsteadiness of the curve is still present, but within 0.1. This test was done just to show the effect of contact pressure on friction. Reducing the contact pressure reduces friction significantly. The variation is still present due to interaction of asperities. The reduced normal pressures would allow more gradual wear of the surface. The contact point showed some sort of flat shining area indicating the initial gross sliding regime of fretting contact.

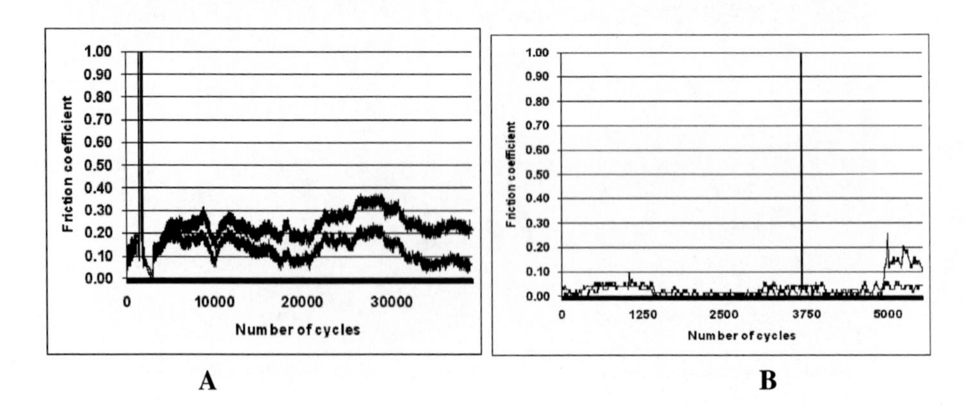

Figure 5.45. Friction curves for contact pressures of (a) 40 MPa, (b) 20 MPa.

5.5.3.3. Plasma Nitriding

Figure 5.46(a) shows fretting damage site of plasma nitrided specimen. The lives of unmodified specimens were less than 18% compared to plasma nitrided specimens under fretting fatigue conditions. It is observed that the surface is not uniformly covered with oxides which mean that proper contact may not always occur between the mating components during fretting. Nitrided layer being stronger than unmodified alloys does not allow early damage and oxidation. This delayed phenomenon is responsible for longer life of the coated components. The transformation of alloy to oxides can only happen while the debris particulate dwells within the contact undergoing attrition from the rubbing materials which later gets compacted within the damage location. Both the pads and the specimen almost suffer the same form of damage. Figure 5.46(b) shows another contact area with three body wear phenomena. The surface has been plowed by the release of few hard particles from either of the mating surface. The particle is released from combined influence of tangential force and normal pressure initiating cracks within the composite modified layer. The three different phases present in the layer are TiN, Ti_2N and matrix.

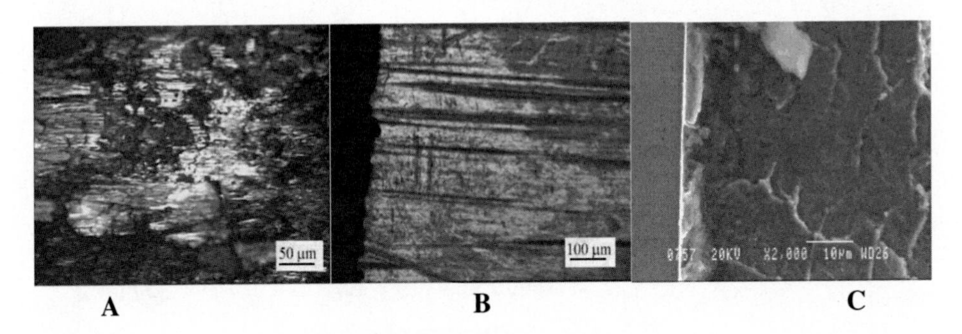

Figure 5.46. (a) Oxidative fretting damage, (b) Three body wear during fretting by release of particulate from the mating contacts, (c) Fractography of plasma nitrided Ti-6Al-4V at 500 MPa.

It would mostly initiate at the junction of the same phase or different phases which normally have different mechanical properties. All of them have variations in strength and hardness level. The failure was along the edge of contact and increase in pad constraint due to presence of released particle may be responsible for the failure. The surface is still softer compared to PVD TiN layer (Nano indentation results) and hence could accommodate higher strains compared to the former case. The gradient layer can offer higher resistance to final failure than surface coatings with abrupt change of properties along the interface. Oxidation was also observed at some contact points.

Figure 5.46(c) shows the cross section of the fractured specimen. The ruptured gradient layer is also visible in the micrograph. Sharp interface cannot be expected due to the nature of the process which involves diffusion of nitrogen within the surface. Faceted feature in the interior of the specimen indicates the plane strain propagation of crack till the ultimate fracture of the specimen. Beach marks and fatigue striations were not observed. Plain fatigue loading will normally initiate subsurface cracks. Fretting induces surface defects deliberately from the surface. Figure 5.47 shows the friction coefficient curve for axial load of 500 MPa. It can be seen that friction coefficient is well within 0.1 throughout the test except at the time of failure.

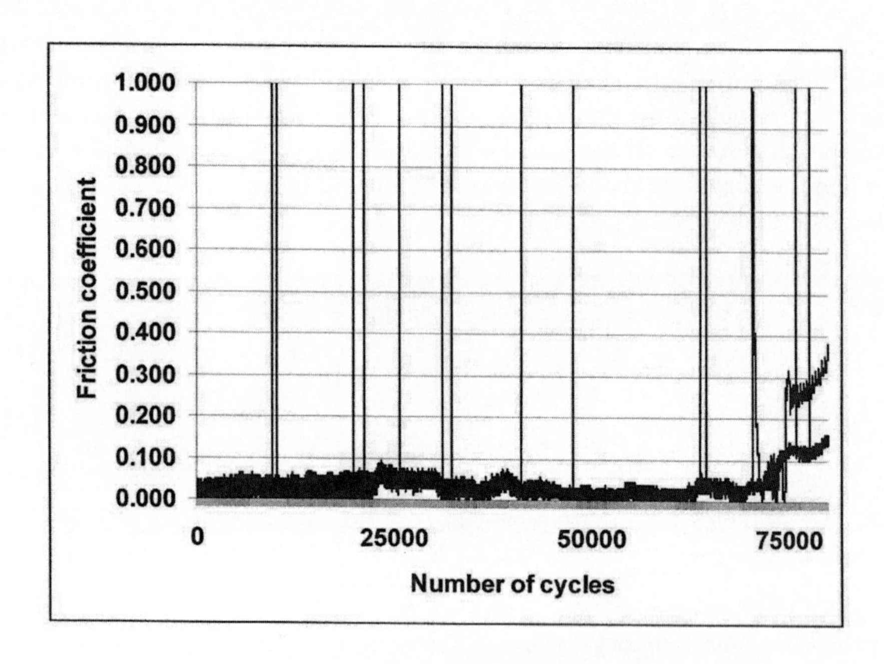

Figure 5.47. Friction curves for plasma nitrided samples.

High friction at the end can be attributed to third body interaction within the contact which may have come into effect after 60,000 cycles with ejection of coarse particles. The failure, as explained earlier, is due to inability of the coating to accommodate the surface strains effecting from increased constraint of pads due to release of nitride particle. The contact area being smoother must normally produce high friction. But the high hardness and chemical inertness combined with lubrication offered by the ringer fluid has reduced adhesion and friction to a low level which is very favorable for implants in minimizing fretting debris generation and surface damage.

5.5.3.4. Ion Implantation

Figure 5.48(a) shows the fretting damage area of ion implanted Ti-6Al-4V at 500 MPa axial stresses. The mode of damage is almost similar to unmodified specimens with lumps of oxides. Thinner implanted layer has worn off within few cycles exposing the bare surface for deeper damage. In this case, we may infer that the dose of 2×10^{16} ions/cm^2 can only offer little improvement in fretting fatigue strength. In general, higher dose of implantation produces higher resistance to wear damage from the formation of nitride precipitates. The beam line mode of ion implantation has specific disadvantage of its inability to modify complex geometries and hence it is time consuming and costly.

Figure 5.48(b) shows the cross section of the fractured specimen indicating no sign of surface layer. At the edge it is seen that crack has taken an inclined path in the beginning and later aligned perpendicular to the stress axis which is indicative of typical feature of fretting fatigue crack propagation as mentioned earlier. The modified layer being very thin has not shown any sign of brittle feature at the surface. Internal surface shows some kind of any oxidized debris formed during gradual crack propagation.

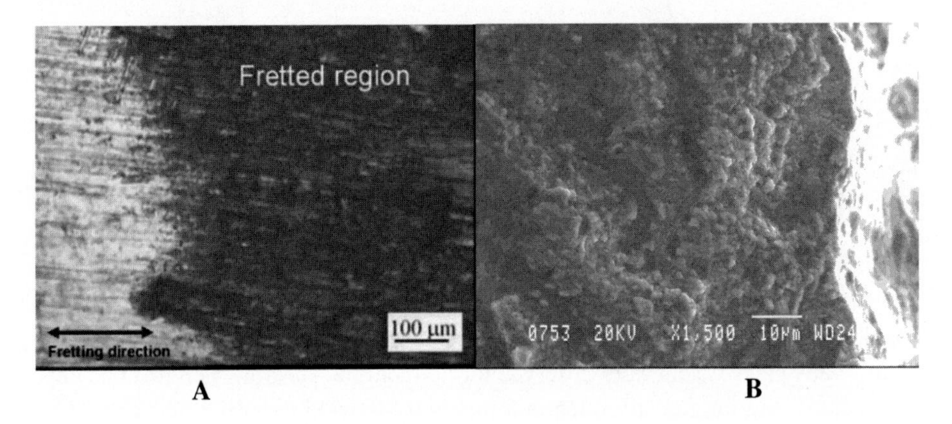

Figure 5.48. (a) Fretting fatigue damage (b) Fractography of ion implanted Ti-6Al-4V at 500 MPa

Figure 5.49 shows the friction coefficient curves for fretting fatigue of ion implanted Ti-6Al-4V alloy. Two curves indicate the average friction of the two pairs of pads on either sides of the specimen. It is observed that friction is high during the intermediate period which may be attributed to the contact of bare surfaces. The protection of ion implanted layer may have been before this period when the friction is low. Unlike the friction curve of unmodified specimens which has shown sudden increment to a high value just after few cycles, the ion implanted layer has protected the material to some extent. Ion implanted layer is very thin and easily wears off within few cycles and the exposed bare surface comes in contact affecting the friction in that region. It is almost similar to unmodified alloys where the friction increases to 0.6 and gradually drops to 0.3. Second pad has maintained the same increase and decrease of friction, but within the lower regime of 0.3. The critical damage may have occurred during the intermediate period when the friction is high. The effect of biofluid and oxide bed is also considered in reducing friction similar to fretting of unmodified alloy.

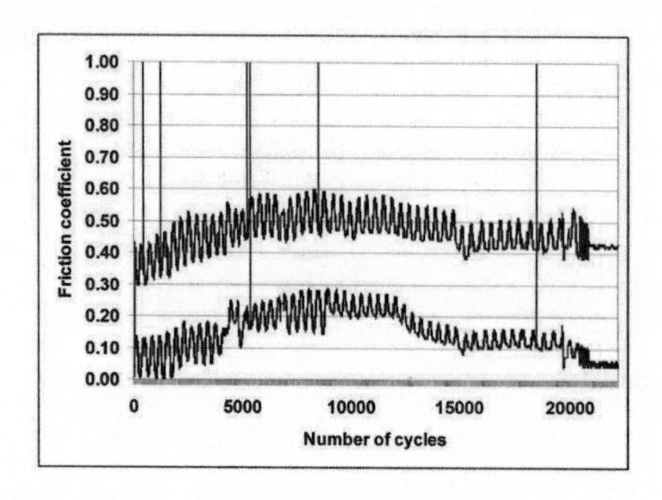

Figure 5.49. Friction curve of Ion Implanted Ti-6Al-4V.

5.5.3.5. Laser Nitriding

Figure 5.50(a) shows fretting fatigue failure edge of laser nitrided Ti-6Al-4V specimen. Cracks have developed perpendicular to the loading axis at the fretting edge. Many such cracks were observed through out the specimen length. The gage thickness of the fatigue specimen is 5 mm and the weld deposit along with HAZ occupies about 1 mm on both the sides. 2 mm out of 5mm thickness constituted hard and brittle phase (TiN dendrites). The axial load applied must have crossed the yield strength of the melt-HAZ composite which will not allow any plastic flow of material.

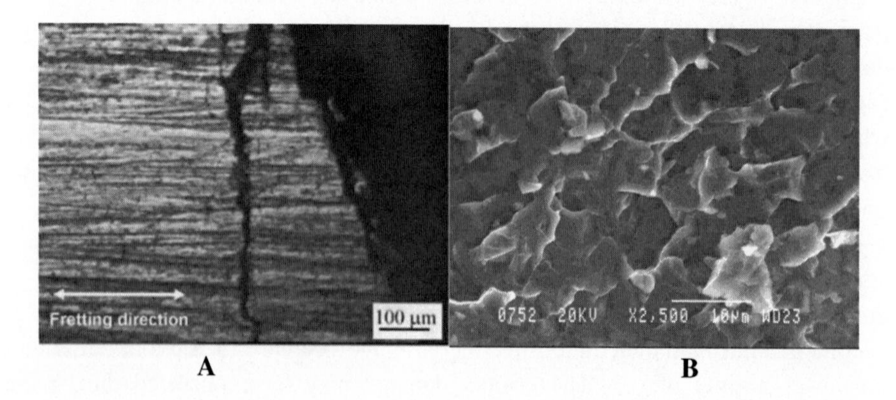

Figure 5.50. (a) Fretting fatigue failure edge (b) Fractography of laser nitrided Ti-6Al-4V specimen at 334 MPa.

Figure 5.50(b) shows cross section of the nitrided portion of the failed sample with the faceted features indicating brittle mode of fracture with no gross plastic deformation.

Figure 5.51 shows networks of cracks on the nitrided region of the grip portion of the specimen developed during fretting fatigue. They are almost perpendicular to the loading direction (shown with arrow). It is clearly due to inability of the nitrided portion to accommodate the axial strain during cyclic loading. The same cracks are not observed in the substrate portion above the nitrided region of the specimen because of higher ductility and strain tolerating capacity of the material. The process of laser nitriding requires much improvement before it can be adopted for surface engineering of bioimplants. It may be difficult to obtain efficient nitriding if the melted zone is thin because it has a larger volume of unmelted substrate below the surface for quicker removal of heat. In such cases, thicker nitride case is unavoidable and the results in tensile loading will be the way as shown here.

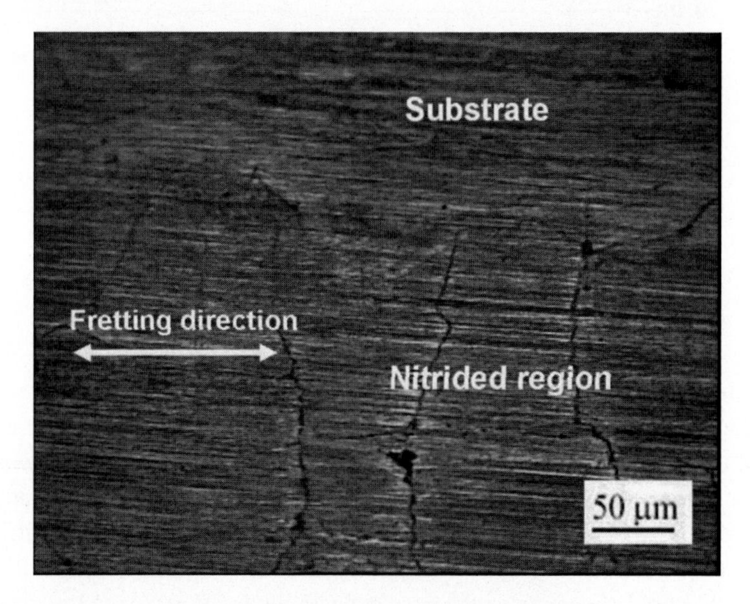

Figure 5.51. Crack networks on the specimen grip portion developed perpendicular to the loading axis (arrow indicates the direction of loading).

Figure 5.52 shows the friction curve for fretting fatigue of laser nitrided titanium alloys. The friction coefficient nearly touches 0.1 and shows a little increment before final fracture. Hard surface asperities have a role to play in inducing friction by interlocking. The case is very hard and chemically inert by nature. It is shown to have low resistance to sliding motion.

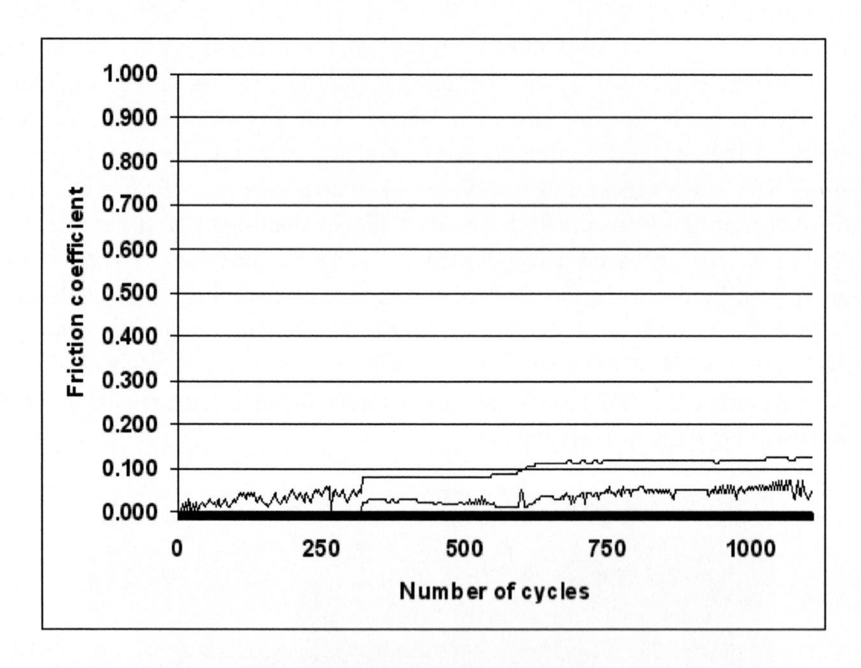

Figure 5.52. Friction curve for fretting fatigue of laser nitrided Ti-6Al-4V alloy.

5.5.3.6. Thermal Oxidation

Figure 5.53(a) shows the fretting fatigue failure edge of the fatigue specimen indicating the brittle failure of the oxidized specimen. Some part of the case has chipped off during fretting oscillation as seen in Figure 5.53(a). As seen from the cross section of the oxidized specimen earlier, the growth of oxide layer is very uneven during the process of thermal oxidation and hence it was difficult to remove the black oxide layer completely. An attempt to remove the layer completely would also damage the case beneath the layer exposing the underlying substrate. Discontinuous film may not be acceptable for load bearing implant applications. The failure must have initiated along such discontinuities. Black oxides were not observed in the failed specimens along any of the contact points. Figure 5.53(b) shows the cross section of the fractured oxidized surface. It is similar to laser nitrided surface which shows faceted features indicative of brittle failure.

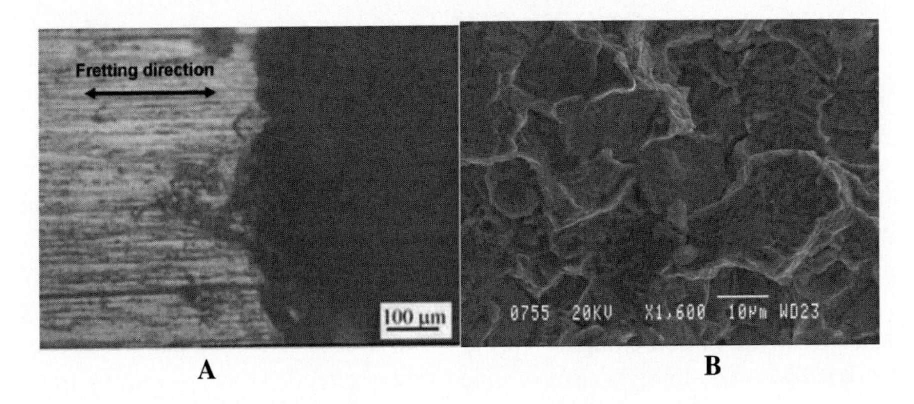

Figure 5.53. (a) Fretting fatigue failure edge (b) Fractography of thermally oxidized Ti-6Al-4V specimen at 400 MPa.

Figure. 5.54 shows the friction curve for fretting fatigue of thermally oxidized titanium alloy pairs. The specimen fractured in very short time, but the friction experienced by the surface is very low. The value is well below 0.03. No other coatings discussed earlier have experienced such low friction between the contact pairs. Therefore the failure is mainly due to irregular (wavy) modified case. Since it is difficult to obtain uniform case depth, thicker case was selected, but the purpose is not served due to weakness of the case itself. The specimens cannot be prepared properly without damaging the case as discussed in fretting wear experiments. The process should be carefully controlled to obtain uniform case. Controlled atmospheres can help achieve better uniformity in the case.

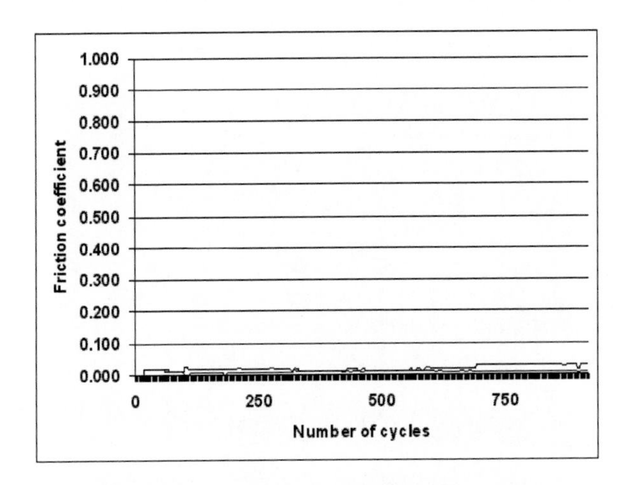

Figure 5.54. Friction curve of thermally oxidized Ti-6AL-4V alloy.

5.6. Some Important Features of Fretting Fatigue Failures

In this section, general features of fretting fatigue failures are explained with some analysis on crack propagation mode during fretting fatigue, the importance of contact configuration following a general model on crack nucleation in unmodified and surface modified pairs.

Figure 5.55 shows the fretting fatigue failure site of a sample. It clearly indicates the fact that fretting failures are always expected at the pad – specimen contact region due to higher stress concentration effecting from severe frictional damage due to fretting oscillation under contact pressure. The sliding motion has also produced visible tracks on the surface as observed in Figure 5.55 indicated by an arrow. The damage slowly compounds with time and nucleates numerous cracks of varied sizes within the contact area as shown in Figure 5.56. From there on, the axial load takes charge for the critical crack to propagate to final fracture.

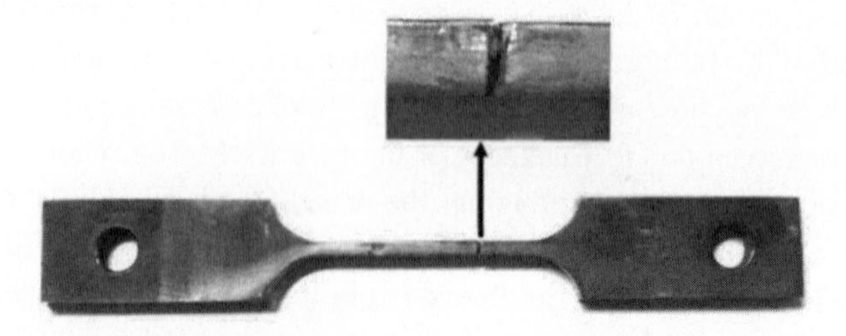

Figure 5.55. Fractured site of a fretting fatigue test specimen.

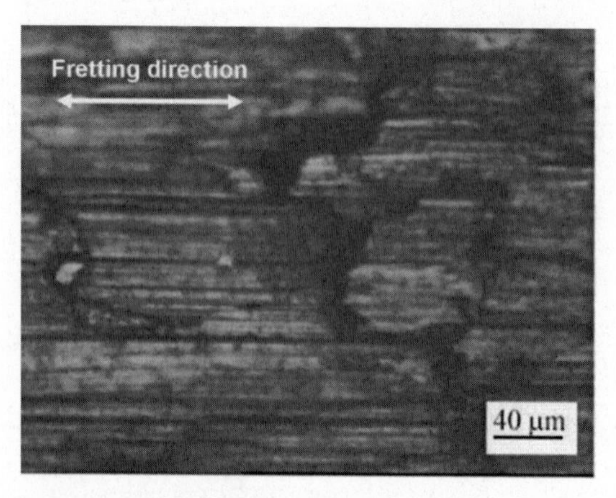

Figure 5.56. Cracks within the fretted region.

The crack usually initiates along the edge of contact (EOC) or in between the stick-slip boundary.

The cracks will initially propagate inclined to the axial load and later detours perpendicular to the load axis when the effect of tangential force diminish below the surface as seen in Figure 5.57. It is explained that the crack will generally propagate along some weak planes almost 45 degrees to the load axis. This is generally considered due to effect of tangential force on the surface. Some of these primary cracks are also responsible for releasing debris particles which will oxidize during later part of cycle or gets compacted within the oxide bed already present. The microscopic examination of the damage site in fretting fatigue tests often reveals what is called tribologically transformed structures (TTS) (Sauger et. al., 2000), as shown in Figure 5.58(a). The black splotch of rutile (TiO_2) has formed during fretting fatigue test without biofluid as confirmed from the XRD analysis in Figure 5.58(b). It may not form to this extent due to limited availability of oxygen within the body. But the physiological or biological fluid surrounding the contact area may influence different tribo-chemical reactions. It is confirmed that the temperature at the contact area often reaches very high level to cause transformation of structures in contact materials (Antoniou and Radtke, 1997). Rutile formation is often experienced during oxidation tests of titanium alloys. Therefore it can be inferred that the contact area has experienced high temperature during sliding motion.

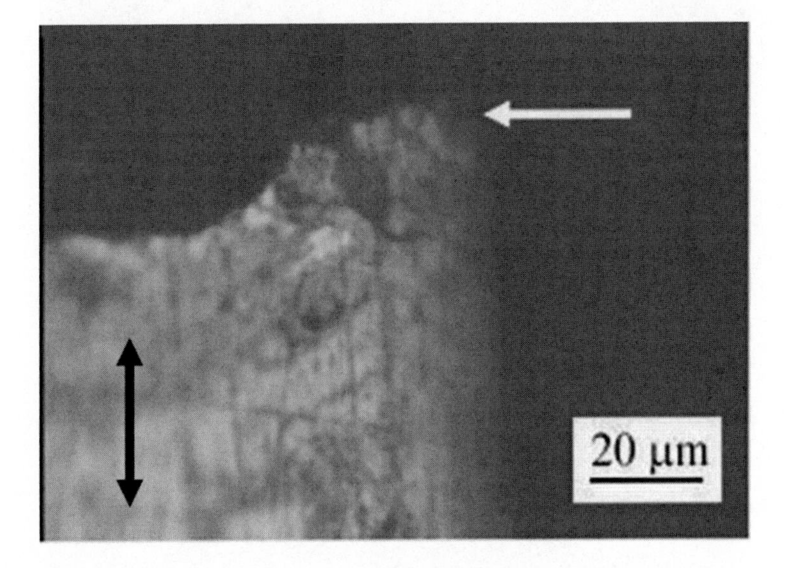

Figure 5.57. Cracks inclined to tensile axis after initiation during fretting fatigue.

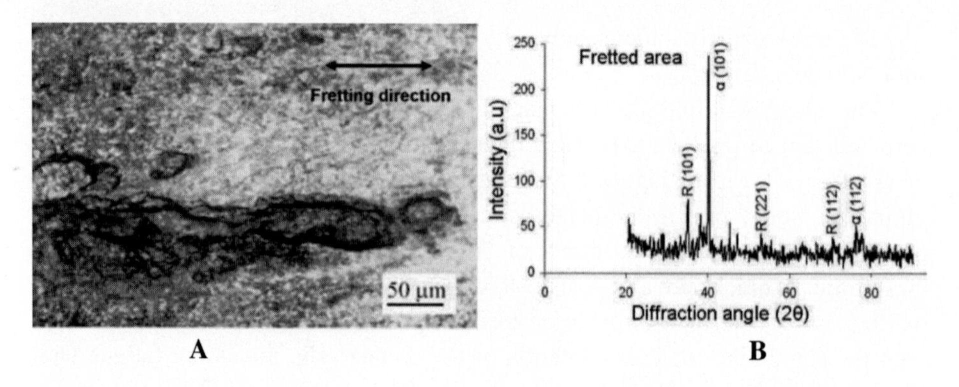

Figure 5.58. (a) Rutile splotch at the fretted area (500X), (b) XRD of the fretted area.

The tissues may not be able to tolerate the chemical effect of rutile and the damage caused may also introduce favorable sites for stress concentration and early crack initiation and propagation due to fatigue loads in hip joints. The fretting oscillation also depends upon the design of the prostheses or implants. Under severe load excursions, as in the case of climbing staircase etc., it may experience severe deflection in case of hip joints and fretting action is inevitable.

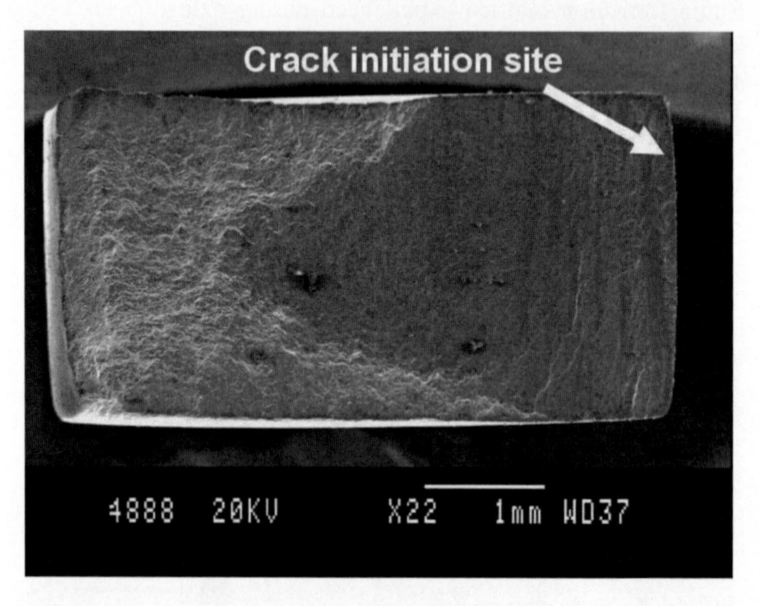

Figure 5.59. Cross section of fractured fretting fatigue test specimen.

The cross section of the fractured specimen is also shown in Figure 5.59. The crack has been initiated from the right side of the specimen cross section as shown in figure and gradually propagated inward (towards left of the specimen). Closer observation in SEM did not reveal any striation marks that are often the notable feature of the fractured cross sectional surfaces of fatigue samples. Some literatures reports striation marks in case studies of failed implants (Teoh, 2000). The flat area in Figure 5.61 indicates that fatigue crack has propagated in a plain strain condition with repeated pounding of the surface against each other.

When the cross section was no longer able to bear the load, the specimen separated in a ductile manner as observed in left region of the specimen. Closer examination in left area revealed dimpled surface with a network of teared ligaments which is a typical feature of a quasi-cleavage mode of fracture. Higher axial load increase the area with quasi-cleavage features and lower axial load increases the plain strain area. The growth of fatigue crack is slower when the axial load is reduced. EDS analysis in this area also shows sodium and chlorine peak, possibly entered from the saline fluid used in these tests. Corrosive conditions are often aggravated from oxygen depletion and chlorine ion concentration around the damage area.

5.7. Fretting Fatigue Process

This description is a general analysis of mechanism of fretting fatigue damage based on the results obtained from the above tests. When two surface of the solid are bought together as shown in Figure 5.60, the real contact is actually at the projected asperities of both the surface which is only few percent of the actual area of the surface. The actual area is determined by a range of material properties such as compressive yield strength and surface characteristics such as roughness and surface films. The real contact would tend to increase with the sliding motion under the combined action of normal and tangential forces that acts on the surface. During sliding of the bare surface, the junction would microweld to form what is called super junctions. Sliding of chemically inert surface would produce a different effect. High friction is produced during formation of super junctions and breakaway of junction would release the surface and produces surface damages and debris particles. With the growth in junctions, the shear force required to cause sliding will increase with each cycle.

In Figure 5.60(a), when surface A comes in contact with surface B, it may cause shear force to act within the asperities (A or B) and subsequent transfer of ejected particles or cracks may begin develop and grow at points A1 and A2 or B1 or B2 or a combination of both. The crack may increase in length due to action of friction and tensile forces and propagate to failure. Due to large number of such asperities, number of cracks would nucleate at the fretted area and finally link up to a critical crack. The contact of biofluid within the cracks may tend to dissolve fresh, uncontaminated and active surface atoms to form chlorides or some other complex compounds. Corrosion products also tend to affect the stress concentration within the cracks.

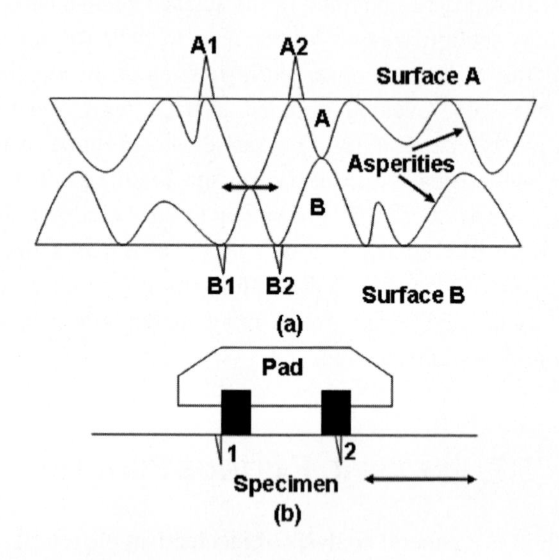

Figure 5.60. Crack nucleation model in fretting fatigue test.

Corrosive attack without stress often forms pits on the alloy surface. These pits acts as notches and reduce the general fatigue strength. The effect of fretting is much more deleterious because it aids corrosion more easily by repeated exposure of fresh surface by deformation and delamination.

The propagation of crack is modified due to presence of transformation products. Oxide induced crack closures is the possible mechanism for reducing the effective stress concentration at the crack tip. So the removal of oxides from the fretted area may reveal such cracks. In Figure 5.60(b), crack 1 tends to behave differently from crack 2 because surface frictional forces tend to open type 1 crack. Type 1 is more deleterious than 2 because crack closure mechanism may become active within type 2 cracks to reduce the effective stress concentration. Crack extension become more prominent in type 1.

Fretting fatigue damage sequence of surface modified alloys can be apparently explained as shown in Figure 5.61. Initial sliding motion between hard and chemically inert surfaces will naturally produce low friction if the roughness level is within certain limits. Uneven surface profile will produce high initial tangential force due to interaction from surface asperities until they fracture when the force crosses elastic limit. Then the sliding motion steadily produces wear particles which get crammed within the contact and increases damage severity as shown in Figure 5.61(b). The wear particles along the trailing edge of the pads are ejected out of the contact, but the middle ones are entrapped within the region and further dig into the surface because of high contact pressure. Then the contact pressure may slightly reduce, but the fretting motion has already initiated the crack. Contact experience frequent shift in the fretting regime (gross slip, partial slip and stick) by oxidation and corrosion reaction as shown in Figure 5.61(c).

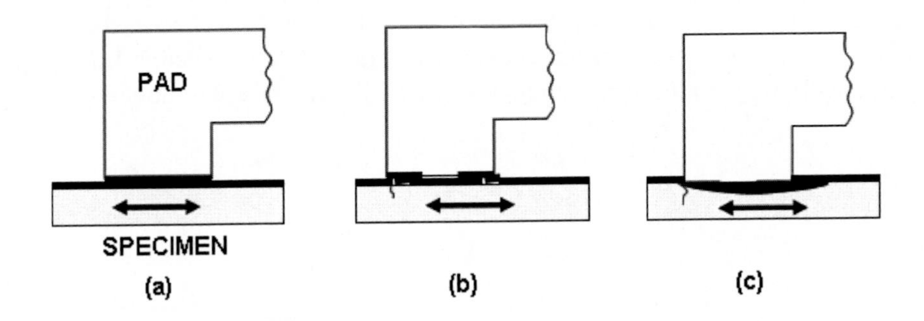

Figure 5.61. Fretting fatigue damage mechanism sequence for surface modified alloys.

The delaminated particles are found squashed within the compacted oxide. Within the body environment, there may be a different tribochemical reaction at the contact leading to formation of different kind of products. In case of unmodified alloys, the initiation of crack usually begins along one of the weak slip planes inclined to the load axis. But for surface modified alloys, it depends upon the physical nature and properties of the layer. Hard ceramic layers do not generally undergo any gross deformations like metallic systems. So when the tangential force crosses the elastic limit of the layer, they tend to rupture. The crack is usually initiated at the edge of contact because of the constraint offered by the pad at that boundary. Variation in contact pressure due to formation of debris could be one of the reason for pads to offer increasing constrain leading to brittle rupture.

Surface modified layers normally offer better resistance to than surface coatings due to absence of interface delamination effecting from stress concentration during fretting action for coatings. Shear deformation at the interface of surface coating depends upon the adhesion strength of the layer on the substrate. Where as the modified layers does not undergo any mechanical detachment like coatings. There is no sharp interface between the substrate and modified layers which shows diffused surface profiles. Therefore modified layers may require higher shear stress to break the particle from the substrate.

Fretting damage is understood to originate from myriad of interacting mechanical and chemical phenomenon like wear, corrosion and fatigue. So the all inclusive model of crack nucleation will lead us to an absurd region. Even in highly controlled experiments it is very difficult to predict the origin and orientation of fretting cracks. Therefore explanation of the fretting phenomenon merely by observation of cracks has deluded many of the researchers. There are many semi-empirical models based on contact-stress field and surface slip amplitude to interpret these observations. Many fretting researchers believe that fretting is a gradual phenomenon with damage gradually accumulating leading to crack nucleation. So it can be termed as crack nucleation instead of initiation.

Chapter 6

CONCLUSION

6.1. PREFACE

Fretting failures of biomedical titanium alloys used for hip prostheses has been an important area in the field of orthopedic surgery. The substrate materials (Ti-6Al-4V and Ti-6Al-7Nb) used for this work represents one of the possible pairs of materials in contact during fretting along modular junction of hip joints or bone plate-screw combination. The surface treatments given are plasma nitriding, PVD TiN coating, ion implantation, laser nitriding and thermal oxidation. Each one of the surface layers are individually characterized with optical and SEM microscopes, X-ray diffraction analysis, roughness profile, microhardness, nano indentation and scratch test. The surface damage during fretting of individual coatings has been presented with both optical and SEM micrographs and the damage sequence have been explained with friction coefficient curves. Fretting wear and fretting fatigue process cannot be compared due to different couplings. Fretting wear behavior is studied with alumina ball on different modified layers, where as fretting fatigue behavior is studied with couplings made of same modified materials. The results of these experiments give some understanding of the damage behavior of surface modified implants within the body. The following text explains the conclusive remarks of all the tests conducted.

6.2. Characterization of Surface Coatings and Modified Layers

In this work, five different coatings were chosen for the surface modification of titanium alloys. Substrate titanium alloys have the hardness of around 380 VHN (4 GPa).

PVD TiN coating is a mechanically bonded film of 2μm thickness with the hardness and reduced elastic modulus of 40 GPa and 334 GPa respectively. The adhesion strength of the coating is 90N. XRD analysis shows pure TiN phase on the substrate.

Plasma nitriding is a surface modification process with diffusion of nitrogen under the effect of thermal and electrical potential. XRD data shows TiN and Ti_2N phases within the layer (3μm). Gradient hardness profile indicates gradient nitrogen concentration within the layer. Nano indentation test shows the hardness and elastic modulus of the layer is 23 GPa and 201 GPa respectively.

Ion implanted sample is 170 nm thick and shows little improvement in properties over the substrate for the dose of 2×10^{16} ions/cm^2. The hardness and elastic modulus is 6 and 115 GPa respectively. It shows no contrasting change in the surface color after implantation.

Laser nitriding of titanium alloys produces super hard TiN dendrites (60 – 70 Vol. %) within the melted region. This has posed difficulty in surface preparation. The modified case is very thick (~0.5mm) with hardness varying between 1450 – 1600 VHN. XRD data shows TiN and Ti_2N in the modified layer.

Thermal oxidation produces black rutile layer above the hard α case (solid solution of oxygen in titanium). It also shows gradient hardness below the surface with the hardness of around 1100 VHN. The case developed by thermal oxidation is highly irregular with deeper and shallow oxidation at different regions.

6.3. Fretting Wear Test Results

The following section gives the test results of fretting wear conducted with alumina ball on flat specimen configuration. In every case, the damage is a localized irregular or regular scar on the surface.

6.3.1. Unmodified Alloys

Unmodified alloys have shown irregular scar in air and circular scar in ringer fluid. The friction coefficient varies in the range of 0.8 to 1.0 in both cases. Rutile oxides are observed in scar tested in air and absent in case of scar tested in ringer fluid. The scar is wider and deeper in ringer fluid than air. This is due to electrochemical influence during fretting ringer fluid.

6.3.2. PVD Tin Coating

The scar pattern for PVD TiN coated alloy is irregular in both air and ringer fluid. The friction coefficient is below 0.2 in air as well as ringer fluid. Oxidation is observed in air, but not in ringer fluid. The scar width is shallow and irregular in both cases compared to unmodified alloys. This indicates that TiN layer is more resistant to mechanical abrasion as well as electrochemical oxidation.

6.3.3. Plasma Nitriding

Plasma nitrided layer is softer than PVD TiN layer and the scar profile is similar to unmodified alloys after the fretting wear test. Compacted oxides are observed after the test in air and scar width is deeper and wider in ringer fluid without oxides. Friction coefficient ranges from 0.4 to 0.7 in air and 0.2 to 0.6 in ringer fluid.

6.3.4. Ion Implantation

Ion implanted layer is more softer and thinner compared to plasma nitrided and the scar profile is as good as uncoated alloys with compacted oxides in air and deep and wide scar in ringer fluid without oxides. Electrochemical oxidation has influenced during fretting in ringer fluid. Friction coefficient ranges from 0.3 to 0.7 during test in air and 0.2 to 0.4 in ringer fluid.

6.3.5. Laser Nitriding

Laser nitrided layer is very thick and hardest of all the coatings. Fretting scar is shallow and irregular in both air and ringer fluid. It is attributed to the presence of large volume percent of closely spaced TiN dendrites in the layer. Even though damage is minimal, friction coefficient is quite high. It ranges between 0.6 to 0.7 in air and 0.3 to 0.4 in ringer fluid.

6.3.6. Thermal Oxidation

Thermally oxidized layer is also thick, but less hard compared to laser nitriding and PVD TiN coating. The wear scar is irregular in both air and ringer solution. The depth and width of the scar is intermediate between uncoated and laser nitrided alloys. Friction coefficient ranged from 0.4 to 0.7 in air and 0.4 to 0.6 in ringer fluid.

6.3.7. Summary of Fretting Wear Results

In the present study on fretting wear of surface modified titanium alloys, Laser nitrided and PVD TiN coated alloys have shown the best performance due to large volume of hard and higher wear resistant TiN phase. The force-displacement loop area is also less for PVD TiN layer indicating the superior wear resistant nature of the layer. The wear volume generated during fretting is minimal primarily due to TiN. Thermally oxidized alloy has shown intermediate performance and plasma nitrided and ion implanted alloys have shown very poor performance due to softer and thinner layers. In terms of quantitative results, the average wear rates for laser nitrided, PVD TiN coated and thermally oxidized samples is 0.6, 3 and 12.5% respectively compared to ion implanted and plasma nitrided samples having shallow thickness and low hardness. Fretting wear rate is higher in uncoated, ion implanted and thermally oxidized alloys due to fretting assisted electrochemical dissolution process.

6.4. Fretting Fatigue Test Results

The following section describes conclusive remarks on the performance of surface modified alloys under fretting fatigue conditions.

6.4.1. S-N Curve

Fretting fatigue has reduced the life of uncoated specimens compared to surface modified specimens. For example, Fretting fatigue life of uncoated pairs is 15 to 18% of plasma nitrided pairs, 45 to 50% of PVD TiN coated pairs and 60 to 65% of the ion implanted pairs at 500 MPa axial stresses. The difference and scatter in results is more at lower axial stresses. Plasma nitrided pairs have shown the best results over all the coatings due to gradient hardness from the surface. Laser nitrided and thermally oxidized pairs have undergone premature failure due to irregular case thickness.

6.4.2. Slip Amplitude

The slip between pad and specimen increases with increase in the applied stress. The slip is slightly high in ringer fluid due to release in constraint from the flow of ringer fluid. During cyclic loading, large variations are observed due to frequent shift in fretting regime (Gross slip, partial slip and stick).

6.4.3. Surface Damage Characterization

6.4.3.1. Unmodified Alloys

Fretting of unmodified alloys shows intense oxidation and surface damage due to high metallurgical compatibility of materials which induces frequent adhesion and rupture of exposed surface. Formation of rutile along the fretted area is confirmed with XRD analysis. Friction curves indicate the mode of contact behavior during fretting. Friction coefficient has declined gradually due to lubrication effect from biofluid and oxide debris bed within the contact. Specimen will fail when the cross section will no longer endure the axial stress.

6.4.3.2. PVD Tin Coating

PVD TiN coating is characterized by formation of hard TiN layer on the surface with sharply visible interface. XRD analysis confirms the formation of pure TiN phase on the surface. Plain fatigue of TiN coated Ti-6Al-4V did not fail even after million cycles because the surface layer prevented the formation of extrusions and intrusions. Fretting fatigue damage is characterized by delamination and compaction of TiN particles within the oxide bed. The life of all the specimens with TiN layer has improved compared to unmodified alloys indicating the positive effect of coating. Friction coefficient curve indicate the unsteady nature of the contact during fretting. But friction coefficient is below 0.4 throughout the test compared to uncoated alloys. Repeated interaction of surface asperities and rupture during fretting can be understood from the initial part of friction coefficient curve. The uneven nature of the curve can be attributed to complex interaction between roughness peaks, debris particles and lubricating effect of biofluid. So fretting is mostly on the surface of the coating before delamination. High friction during the later part of the cycle explains the failure of coating. The effect of biofluid and oxide bed in reducing the friction is also confirmed from the last part of the friction curve.

6.4.3.3. Plasma Nitriding

Plasma nitrided surface is smoother than as polished unmodified alloys. XRD analysis confirms the formation TiN and Ti_2N capable of resisting wear damage. Surface modified layers are much better than coatings due to absence of sharp interface between substrate and surface modified layer. However layers of higher thickness will have negative effect on fretting fatigue life. Fretting fatigue damage area of plasma nitrided titanium pairs is also characterized with oxidation and deep scratches effecting from third body wear mode of contact interaction. Friction coefficient remained below 0.1 for most of the cycle. The surface damage is very gradual.

6.4.3.4. Nitrogen Ion Implantation

Nitrogen Ion implanted layer shows no visible changes or features on the surface. Even though it is a surface modified layer, it can perform better only under low normal load due to low thickness. Implanted depth was reported as 170 nm from the TRIM software which can calculate the projected range of ions for a material with particular combination of elements. The properties of the layer can be well characterized with nano indentation. Scratch test can only indicate the presence of hard modified layer. Fretting fatigue induced damage is almost comparable with unmodified alloys having large amount of oxides within the damaged area. Friction curve shows gradual increase and decrease with the number of cycles. Increase in friction during first half of the life is attributed to slow damage of surface layer and adhesion of surface. Decrease in friction during later part of the cycle before failure is attributed to lubricating effect of biofluid and oxide debris.

6.4.3.5. Laser Nitriding

Laser nitrided titanium alloys contain large volume fraction of TiN dendrites within the melt deposit on the surface. The melt deposit itself is almost 0.5mm thick from the surface. All the specimens experienced brittle failure during fretting fatigue test within short cycles due to hard and thick modified layer. Friction coefficient is below 0.1 till final failure. The separation of the specimens was at the contact edge and some of them failed away from the contact area due to thick and brittle modified layer. Oxidation was not observed in any of the specimens. The process of laser nitriding needs improvement for bioimplant applications. It cannot be used for implant with complex geometries and also not usable for mass production. Post operative machining and further surface preparations may be necessary which generally pose difficulties.

6.4.3.6. Thermal Oxidation

Thermally oxidized surface is characterized by hard α case below the layer of rutile. Oxidation of titanium alloys at high temperature allows thermally aided diffusion of oxygen forming non uniform hard case. It is reported to have excellent tribocorrosion and wear characteristics. But fretting fatigue life is very poor due to non-uniform case thickness. The process of thermal oxidation used in this work is not very suitable for fatigue applications. Like laser nitrided alloys, many specimens also failed within short number of cycles. Friction coefficient was below 0.05 throughout the test for one of the specimen tested at 334 MPa. Fretting damage and oxidation was not observed in any of the specimens tested. There seems to be sharp fatigue limit above which the specimen cannot endure and below which it would never fail. Surface preparation after thermal oxidation is difficult due to uneven nature of oxide growth.

6.4.3.7. Summary Results of Fretting Fatigue Tests

In the present study on fretting fatigue behavior of surface modified titanium alloys, plasma nitriding process has shown the best performance in terms of increasing fretting fatigue life due to shallow gradient layer having higher capacity for accommodating fretting induced surface deformations thereby minimizing the probability of early crack nucleation. Fretting wear performance is therefore poor. PVD TiN coating is slightly inferior due to mechanical adhesion of the super hard layer. Even though fretting wear performance is good, the layer delamination is responsible for failure during fretting fatigue. Ion implanted layer is shallow and soft with fretting fatigue behavior almost similar to uncoated pairs. Laser nitrided and thermally oxidized pairs have undergone premature failure due to irregular case thickness. Fretting fatigue results cannot be correlated with fretting wear results due to different couplings. The results are highly dependent on the relative hardness of the contact pairs as explained earlier.

6.4.4. General Observations of Fretting Fatigue Failures

Fretting fatigue failure is always observed to initiate along the edge of contact between pad and the specimen. Fretting fatigue life is classified into crack initiation life and crack propagation life. Cracks initially traverses inclined and later align perpendicular to the stress axis. This transition occurs when the effect of tangential force gradually diminishes below the surface during crack propagation in proportionate to increase in the effect of axial load. Numerous cracks can be observed within the damaged site filled with fretting debris. Oxidation is inevitable because the contact experiences high temperatures during fretting process. The cross section of the failed specimen usually shows beach marks and features with quasi cleavage mode of fracture.

REFERENCES

[1] Antoniou R.A., Radtke T.C. (1997), *Mechanism of fretting fatigue of Titanium Alloys*, Materials Science and Engineering A, Vol. 237, 229 – 240.

[2] Alves, Jr. C., Guerra Neto C.L.B., Morais G.H.S., Da Silva C.F. and Hajek V. (2005), *Nitriding of titanium disks and industrial dental implants using hollow cathode discharge*, Surface and Coatings Technology, Vol. 194, Issues 2-3, 196-202.

[3] Ani Zhecheva, Wei Sha, Savko, Adrain Long (2004), *Enhancing the microstructure and properties of titanium alloys through nitriding and other surface engineering methods,* Surface and Coatings Technology, Vol. 200, Issue 7, 21 December 2005, Pages 2192-2207.

[4] Animesh Choubey, Bikramjit Basu, R. Balasubramaniam (2004),*Tribological behaviour of Ti-based alloys in simulated body fluid solution at fretting contacts,* Materials Science and Engineering A, Vol. 379, 234–239

[5] Borgioli F., Galvanetto E., Iozzelli F. and Pradelli G. (2005*), Improvement of wear resistance of Ti–6Al–4V alloy by means of thermal oxidation,* Materials Letters, Vol. 59, Issue 17, 2159-2162.

[6] Badisch E., Fontalvo G.A., Stoiber M., Mitterer C. (2003), *Tribological behavior of PACVD TiN coatings in the temperature range up to 500 °C,* Surface and Coatings Technology, Vol. 163 –164, 585–590.

[7] Bhowmick S., Jayaram V., Biswas S.K. (2005), *Deconvolution of fracture properties of TiN films on steels from nanoindentation load–displacement curves*, Acta Materialia, Vol. 53, Issue 8, 2459-2467

[8] Bulatov V. P., Krasny V. A., Schneider Y. G. (1997), *Basics of Machining methods to yield wear and fretting resistive surfaces having regular roughness pattern*, WEAR, Vol. 208, 132 – 137.

[9] Bekir S. Yilbas, Ahemt Z. Sahin, Ahmed Z. Al-Garni, Syed A. M, Zaki Ahmed, Abdulaleem B. J., Muhammed Sami (1996), *Plasma Nitriding of Ti-6Al-4V alloy to improve some tribological properties, Surface and Coatings Technology*, Vol. 80, 287 – 292.

[10] Chan P. C. and Thornley J. C. (2002), *Common features of fretting fatigue cracking in steels,* Practical Failure Analysis, Vol. 6, 85 – 90.

[11] Chen H., Wu P. Q., Xu K. W. (2002), *Comparison of fretting wear of Cr-rich CrN and TiN coatings in air of different relative humidities*, WEAR, Vol. 253, 527 – 532.

[12] Carpene E., Shinn M., Schaaf P. (2005), *Free-electron laser surface processing of titanium in nitrogen atmosphere, Applied Surface Science,* Vol. 247, Issues 1-4, 307-312.

[13] Chu P. K., Chen J. Y., Wang L. P., Huang N. (2002), *Plasma-surface modification of biomaterials, Materials Science and Engineering R,* Vol. 36, Issues 5-6, 143-206.

[14] David W. Hoeppner, Amy M. H. Taylor, and Venkatesan Chandrasekaran (2003), *Fretting fatigue behavior of Titanium alloys", Fretting Fatigue: Advances in Basic Understanding and Applications, Standardization of Fretting Fatigue Test Methods and Equipments,* ASTM STP 1425, 291 – 306.

[15] Dobromirski, J.M. (1992) *Variables of Fretting Processes: Are There 50 of Them?, Standardization of Fretting Fatigue Test Methods and Equipments*, ASTM STP 1159, 60 - 66.

[16] Dong H., Bell T. (2000), *Enhanced wear resistance of titanium surfaces by a new thermal oxidation treatment,* WEAR, Vol. 238, Issue 2, 131-137.

[17] Elder J. E., Tahmburaj R., Patnaik P. C. (1989), *Optimizing ion implantation conditions for improving wear, fatigue and fretting fatigue of Ti-6Al-4V*, Surface Engineering, Vol. 5, No.1, 55 – 75.

[18] Endo K and Goto H (1976), *Inititation and propagation of fretting fatigue cracks*, WEAR, Vol. 38, 311 – 324.

[19] Ettaqi S., Hays V., Hantzpergue J.J., Saindrenan G., Remy J.C. (1998), *Mechanical, structural and tribological properties of titanium nitrided by a pulsed laser,* Surface and Coatings Technology, Vol. 100-101, 428-432.

[20] Geetha M., U. Kamachi Mudali, R.Ashokmani and Baladev Raj (2003) *Corrosion and Microstructural aspects of titanium and its alloys as orthopedic devices*, Corrosion reviews. Vol 21, Nov 2-3, 125 – 159.

[21] Geetha M., Kamachi Mudali U., Pandey N. D., *Asokamani R., Baldev Raj (2004), Microstructural and corrosion evaluation of laser surface nitrided Ti – 13Nb – 13Zr alloy*, Surface Engineering, Vol. 20, No. 1, 68 – 75.

[22] Galliano F., Galvanetto E., Mischler S., Landolt D. (2001), *Tribocorrosion behavior of plasma nitrided Ti–6Al–4V alloy in neutral NaCl solution*, Surface and Coatings Technology, Vol. 145, Issues 1-3, 121-131.

[23] Gokul Lakshmi S., Arivuoli D. 2004), *Tribological behaviour of plasma nitrided Ti-5Al-2Nb-1Ta alloy against UHMWPE, Tribology International*, Vol. 37, Issue 8, 627-631.

[24] Günzel R., Shevshenko N., Matz W., Mücklich A., Celis J. P. (2001), *Structural investigation and wear resistance of submicron TiN coatings obtained by a hybrid plasma immersion ion implantation process,* Surface and Coatings Technology, Vol. 142-144, 978-983.

[25] Hager C.H., Jr., Sanders J.H., Sharma S. (2004), *Characterization of mixed and gross slip fretting wear regimes in Ti6Al4V interfaces at room temperature*, WEAR Vol. 257, 167–180.

[26] Hallab. N. (2001), *Safety of metallic implants*, Journal of Bone Joint Surgery Vol. 83 - A, 428 - 436.

[27] Hallab. N and Joshua J. Jacobs (2003) *Orthopedic Implant Fretting Corrosion,* Corrosion Review, Vol. 21, 183 – 214.

[28] Hyukjae Lee and Shankar Mall (2004), *Effect of dissimilar mating materials and contact force on fretting fatigue behavior of Ti-6Al-4V*, Tribology International, Vol. 37, 35 – 44.

[29] Hasan Guleryuz, Huseyin Cimenoglu (2004), *Effect of thermal oxidation on corrosion and corrosion-wear behavior of Ti-6Al-4V alloy*, Biomaterials, Vol. 25 3325 – 3333.

[30] Hasan Guleryuz, Huseyin Cimenoglu (2005), *Surface modification of a Ti-6Al-4V alloy by thermal oxidation,* Surface and Coating Technology, Vol. 192 (2005) 164 – 170.

[31] Hansson T, Kamaraj M, Mutoh Y, Pettersson B, *High temperature fretting fatigue behavior of an XD γ-base TiAl*, ASTM STP, Vol. 1367.

[32] Hattori T., Nakamuram M., and Watanabe, T (2003), *Improvement of Fretting Fatigue Strength by Using Stress-Release slits", Fretting Fatigue: Advances in Basic Understanding and Applications,* Standardization of Fretting Fatigue Test Methods and Equipments, ASTM STP 1425, 159 – 168.

[33] Huang nan, Chen Yuanru, Cai Guangjun, Lin Chenggang, Wang Zhongguang, Yiao Guo, Su Huehe, Liu Xianghuai, Zhen Zhihong (1996), *Research on the fatigue behavior of titanium based biomaterial coated with titanium nitride film by ion beam enhanced deposition,* Surface and Coatings Technology, Vol. 88, 127 – 131.

[34] Hultman L., Shinn M., Mirkarimi P. B., Barnett S. A. (1994), *Characterization of misfit dislocations in epitaxial (001)-oriented TiN, NbN, VN, and (Ti,Nb)N film heterostructures by transmission electron microscopy,* Journal of Crystal Growth, Vol. 135, 309-317.

[35] Hu C., Baker T. N. (1999), *The importance of preheat before laser nitriding a Ti–6Al–4V alloy,* Materials Science and Engineering A, Vol. 265, Issues 1-2, 268-275.

[36] Hong Liang, Bing Shi, Aaron Fairchild and Timothy Cale (2004), *Applications of plasma coatings in artificial joints: an overview,* Vacuum, Vol. 73, 317-326.

[37] Hoeppner D. W. (1994), Fretting fatigue, ESIS Vol. 18, 3 – 19.

[38] Jin O. and Mall S. (2002), *Effects of independent pad displacement on fretting fatigue behavior of Ti-6Al-4V*, WEAR, Vol. 253, 585 – 596.

[39] Jin-Ki Hong, In-Sup Kim, Chi-Yong Park, Eung-Seon Kim (2005*), Microstructural effects on the fretting wear of Inconel 690 steam generator tube* WEAR, Vol. 259, 349-355.

[40] Kamachi Mudali U., Sridhar T. M., Eliaz N., Baladev Raj (2003) *Failures of stainless steel orthopedic devices – causes and remedies,* Corrosion Reviews, Vol 21, Nos 2-3, 231-267.

[41] Kamachi Mudali U., Sridhar T. M., Baladev Raj (2003), *Corrosion of bio implants,* Sadhana, Vol 28, Parts 3 & 4, 601 – 637.

[42] Khan M. A., Williams R. L., Williams D. F. (1999) "*Conjoint corrosion and wear in titanium alloys",* Biomaterials, Vol. 20, 765 – 772.

[43] Kozo Nakazawa, Norio Maruyama, Takao Hanawa (2003), *Effect of contact pressure on Fretting Fatigue of 316L Stainless steel,* Standardization of Fretting Fatigue Test Methods and Equipments, ASTM STP 1425, 169 – 182.

[44] Leyens C., Peters M. (2004) *Titanium and Titanium alloys*", Ch. 5, WILEY-VCH GmbH & Co. KGaA.

[45] Laure Duisabeau, Pierre Combrade, Bernard Forest (2004), *Environmental effect on fretting of metallic materials for orthopedic implants,* WEAR, Vol. 256 805 – 816.

[46] Mutoh Y., Satoh T., Tanaka K., Tsunoda E. (1989), *Fretting fatigue at elevated temperatures in two steam turbine steels,* Fatigue & Fracture of Engineering Materials and Structures, Vol. 12, 409 – 421.

[47] Man H.C., Zhao N.Q., Cui Z.D. (2005), *Surface morphology of a laser surface nitrided and etched Ti–6Al–4V alloy,* Surface and Coatings Technology, Vol. 192, 341-346.

[48] Mohmad Soib Selamat, Baker T. N., Watson L. M. (2001), *Study of the surface layer formed by the laser processing of Ti–6Al–4V alloy in a dilute nitrogen environment,* Journal of Materials Processing Technology, Vol. 113, 509-515.

[49] Mutoh Y. (1995), *Mechanism of fretting fatigue*, JSME Vol. 38, 405 – 414.

[50] Mitsuo Niinomi (1998) *Mechanical properties of biomedical titanium alloys,* Materials Science and Engineering A, Vol. 243, 231 – 236.

[51] Marc Long and Rack H.J. (1998), Review – *Titanium alloys in total joint Replacement a material science perspective,* Biomaterials Vol. 19, 1621 – 1639.

[52] Man H.C., Zaho N.Q., Cui Z.D. (2005), *Surface morphology of laser nitrided and etched Ti-6Al-4V alloy,* Surface and Coatings Technology, Vol. 192, 241 – 346.

[53] Mishra S.C., Nayak B.B., Mohanty B.C., Mills B. (2003), *Surface nitriding of titanium in arc plasma, Journal of Materials Processing Technology,* Vol. 132, 143-148.

[54] Papakyriacou M., Mayer H., Pypen C., Plenk H., Jr., Stanzl-Tschegg S., (2001) *Influence of loading frequency on high cycle fatigue properties of b.c.c and h.c.p metals,* Materials Science and Engineering A, Vol. 308, 143 – 152.

[55] Prakash B., Richter E., Pattyn H., Celis J.P. (2003), *Ti–B and Ti–B–C coatings deposited by plasma immersion ion implantation and their fretting behavior*, Surface and Coatings Technology, Vol. 173, 150-160.

[56] Rebecca Cortez, Shankar mall, Jeffrey R. Calcaterra (1999), *Investigation of variable amplitude loading on fretting fatigue behavior or Ti-6Al-4V*, International Journal of Fatigue, Vol. 21, 709 – 717.

[57] Ramesh R., Gnanamoorthy R. (2006), *Development of a fretting wear test rig and preliminary studies for understanding the fretting wear properties of steels*, Materials & Design, Vol. 27, 141-146.

[58] Sumita M., Hanawa T., Ohnishi I., Yoneyama T. (1994) *Failure Processes in Biometallic Materials,* Comprehensive Structural Integrity, Vol. 9, 131–167.

[59] Shenhar A. (2000), *Titanium nitride coatings on surgical titanium alloys produced by a powder immersion reaction assisted coating method: residual stresses and fretting behavior*, Surface ad Coatings Technology, Vol. 126, 210 – 218.

[60] Sinha V., Soboyejo W.O. (2001), *An investigation of the effects of colony microstructure on fatigue crack growth rate in Ti-6Al-4V"*, Material Science and Engineering A, Vol. 319 – 321, 607 – 612.

[61] Sittig C, Hahner G., Marti A., Textor M., Spencer N. D. (1999), *The Implant Material, Ti6Al7Nb: surface microstructure, composition and properties*, Journal of Materials Science in Medicine, Vol. 10, 191 – 198.

[62] Starosvetsky D., Gotman I. (2001), *Corrosion behavior of titanium nitride coated Ni-Ti shape memory surgical alloy,* Biomaterials, Vol. 22, 1853-1859.

[63] Shenhar A., Gotman I., Radin S., Ducheyne P., Gutmanas E. Y. (2000), *Titanium nitride coatings on surgical titanium alloys produced by a powder immersion reaction assisted coating method: residual stresses and fretting behavior*, Surface and Coatings Technology, Vol. 126, 210 – 218.

[64] Sung J.H., Kim T.H., Kim S.S. (2001), *Fretting damage of TiN coated zircaloy-4 tube,* WEAR, Vol. 250, 658–664.

[65] Shengli Ma, Kewei Xu, Wanqi Jie (2004), *Wear behavior of the surface of Ti–6Al–4V alloy modified by treating with a pulsed d.c. plasma-duplex process*, Surface and Coatings Technology, Volume 185, Issues 2-3, 22 July 2004, Pages 205-209.

[66] Sobiecki J. R., Wierzcho T., Rudnicki J. (2001), *The influence of glow discharge nitriding, oxynitriding and carbonitriding on surface modification of Ti–1Al–1Mn titanium alloy*, Vacuum, Vol. 64, 41-46.

[67] Sauger E, Fouvry S, Ponsonnet L, Kapsa Ph., Martin J. M, and Vincent L, (2000), *Tribologically transformed structure in fretting*, WEAR, Vol. 245, 39-52.

[68] Teoh S. H. (2000) *Fatigue of biomaterials: a review*, International Journal of Fatigue Vol. 22, 825–837.

[69] Thair L., Kamachi Mudali U., Bhuvaneswaran N., Nair K.G.M., Asokamani R., Baldev Raj (2002) *Nitrogen ion implantation and in vitro corrosion behavior of as-cast Ti-6Al-7Nb alloy,* Corrosion Science, Vol. 44, 2439 – 2457.

[70] Takahito Ohmura, Saburo Matsuoka (2003), *Evaluation of mechanical properties of ceramic coatings on a metal substrate,* Surface and Coatings Technology, Vol. 169 – 170, 728 – 731.

[71] Tan L, Shaw G, Sridharan K, Crone W C (2005), *Effects of oxygen ion implantation on wear behavior of NiTi shape memory alloy,* Mechanics of Materials, Vol. 37, 1059 – 1068.

[72] L. Tan, G. Shaw, K. Sridharan and W.C. Crone (2005), *Effects of oxygen ion implantation on wear behavior of NiTi shape memory alloy,* Mechanics of Materials, Vol. 37, Issue 10, 1059-1068.

[73] Tan L., Crone W.C. (2005), *Effects of methane plasma ion implantation on the microstructure and wear resistance of NiTi shape memory alloys*, Thin Solid Films, Vol. 472, Issues 1-2, 282-290.

[74] Vincent L. Berthier, Y., Godet, M. (1992) *Testing Methods in Fretting fatigue: A Critical Appraisal, Standardization of Fretting Fatigue Test Methods and Equipments*, ASTM STP 1159, 33 - 48.

[75] Vaquila I., Vergara L. I., Passeggi Jr. M.C.G. (1999), R. A. Vidal and J. Ferrón, *Chemical reactions at surfaces: titanium oxidation, Surface and Coatings Technology,* Vol. 122, 67-71.

[76] Wen-Jun Chou, Ge-Peng Yu, Jia-Hong Huang (2002), *Mechanical properties of Tin coatings on 304 stainless steel substrates, Surface and Coatings* Technology, Vol. 149, 7 – 13.

[77] Wen-Jun Chou, Ge-Ping Yu, Jia-Hong Huang (2003), *Effect of heat treatment on the structure and properties of ion-plated TiN films, Surface and Coatings* Technology, Vol. 168, 43–50.

[78] Wiklund U., Hutchings I.M. (2001), *Investigation of surface treatments for galling protection of titanium alloys*, WEAR, Vol. 251, 1034-1041.

[79] Waterhouse R. B (1972), *Fretting Corrosion*, Ch. 7, Pergamon press.

[80] Xuanyong Liu, Paul K. Chu, Chuanxian Ding (2004), *Surface modification of titanium, titanium alloys, and related materials for biomedical applications*, Materials Science and Engineering R, Vol 47, 49 – 121.

[81] Yasuo Ochi, Yohide Kido, Taisuke Akiyama, Takashi Matumura (2003) *Effect of Contact Pad Geometry on Fretting Fatigue Behavior of High Strength Steel", Fretting Fatigue: Advances in Basic Understanding and Applications, Standardization of Fretting Fatigue Test Methods and Equipments,* ASTM STP 1425, 220 – 232.

[82] Yongqing Fu, Nee Lam Loh, Andrew W. Batchelor, Daoxin Liu, Xiaodong Zhu, Jaiwen He, Kewei Xu (1998), *Improvement of fretting wear and fatigue resistance of Ti-6Al-4V by application of several surface treatments and coatings*, Surface and Coatings Technology, Vol. 106, 192 – 197.

[83] Yves Wouters, Alain Galerie and Jean-Pierre Petit (1997), *Thermal oxidation of titanium by water vapour*, Solid State Ionics, Vol. 104, 89-96.

[84] Young-Ho Lee, Hyung-Kyu Kim (2005), *Characteristic of slipping behavior in vibratory wear of a supported tube*, WEAR, Vol. 259, 337-348.

[85] Yoshimitsu Okazakia, Emiko Gotoh (2005), *Comparison of metal release from various metallic biomaterials in vitro,* Biomaterials, Vol. 26, 11-21.

[86] Zenghu Han, Jiawan, Jijun Lao, Geyang Li (2004), *Effect of thickness and substrate on the mechanical properties of hard coatings*, Journal of Coating Technology: Research, Vol 1, No.4, 337 – 341.

INDEX

C

D

E

F

G

H

I

J

K

L

M

N

O

P

R

S

T

U

V

W

X

Y

Z